The Patrick Moore Practical Astronomy Series

T0259673

For further volumes:
http://www.springer.com/series/3192

Observer's Guide to Star Clusters

Mike Inglis

 Springer

Dr. Mike Inglis
Long Island, USA

ISSN 1431-9756
ISBN 978-1-4614-7566-8   ISBN 978-1-4614-7567-5 (eBook)
DOI 10.1007/978-1-4614-7567-5
Springer New York Heidelberg Dordrecht London

Library of Congress Control Number: 2013941476

Printed on acid-free paper

Springer is part of Springer Science+Business Media (www.springer.com)

Dedication

For Arfur, Bynxie, and Charlie

Acknowledgments

I would like to thank the following individuals and organizations, without whom this book would never have seen the light of day (or darkness of night?).

Maury Solomon, Harry Blom, and John Watson at Springer Publishing. Their belief that the book would one day be finished, along with their patience, allowed me to complete the book and deliver it in its final from, acceptable by all. They also forgave me when I started writing on a completely different topic by mistake.

Carina Software, publishers of the *Voyager* planetarium program. This terrific piece of software allowed me to make the star maps in a manner that was dictated by myself and not by the limitations of a program.

Steve Roach, the musician and composer whose music was a constant companion during the lonely days and nights, as the guide was compiled.

I also want to especially thank the following astronomers, who kindly contributed to the guide. They are some of the most knowledgeable and experienced observers around. Their passion for astronomy, and their willingness to share that passion, is admirable. They are:

Dave Eagle (UK)
Phil Harrington (USA)
Michael Hurrell (UK)
Martin Mobberly (UK)
Jim Mullaney (USA)

Thank you everyone!

Long Island, USA   Mike Inglis

About the Author

Dr. Mike Inglis was born in Wales in the UK, but lives and works in the USA, where he is Professor of Astronomy and Astrophysics at the State University of New York. He is a Fellow of the Royal Astronomical Society, A NASA Solar System Ambassador, a Member of the American Astronomical Society, a Member of the Royal Institution, and a Member of the British Astronomical Association. He is the author of many books and research papers including Field Guide to Deep Sky Objects (Springer, 2012, 2nd Edition), An Observer's Guide to Stellar Evolution (Springer, 2003), Astronomy of the Milky Way, Vol. I and II (Springer, 2004), and Astrophysics is Easy (Springer, 2007). He is a Series Editor of several Springer book series for both amateur and professional astronomers. Although a professional astronomer, he is also a lifelong amateur astronomer, and observes the night sky, around the world whenever time, and the sky, permit.

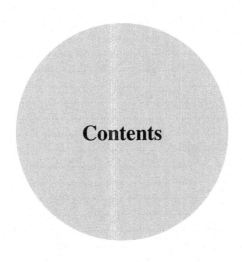

Contents

Introduction to Star Clusters 1
Rationale Behind the Guide 1
Open Clusters 2
   Size 2
   Distribution in the Sky 2
   Appearance 2
   Age 3
   Color 3
Globular Clusters 4
Stellar Associations and Streams 4
Observing Star Clusters 5
   Clothing 5
   Recording Your Observations 6
Classification 6
   Trumpler Classification for Open Star Clusters 6
   Shapley-Sawyer Classification for Globular Star Clusters 7
Constellation Data and Cluster Lists 7
Constellation Maps 9
Constellation List 10
How to Use This Guide 10
And Finally… 11

The Constellations and Their Star Clusters 13
Andromeda 14
Apus 16
Aquarius 18
Aquila 22
Ara 26
Auriga 30
Boötes 34
Camelopardalis 36

Cancer 40
Canes Venatici 42
Canis Major 44
Canis Minor 50
Capricornus 52
Carina 54
Cassiopeia 58
Centaurus 66
Cepheus 72
Circinus 78
Columba 82
Coma Berenices 84
Corona Austrina 88
Crux 90
Cygnus 94
Delphinus 102
Dorado 106
Fornax 112
Gemini 114
Hercules 118
Horologium 122
Hydra 124
Hydrus 128
Lacerta 130
Lepus 134
Libra 136
Lupus 138
Lynx 142
Lyra 144
Mensa 148
Monoceros 150
Musca 158
Norma 162
Octans 166
Ophiuchus 168
Orion 176
Pavo 180
Pegasus 182
Perseus 184
Pictor 190
Puppis 192
Pyxis 202
Sagitta 206
Sagittarius 210
Scorpius 220
Sculptor 230
Scutum 232
Serpens 236
Taurus 240
Telescopium 244

Triangulum Australe 246
Triangulum.. 248
Tucana.. 250
Ursa Major 254
Ursa Minor 256
Vela 258
Virgo.. 264
Volans.. 266
Vulpecula 268

Appendix: Books, Magazines, and Organizations 273

Index.. 275

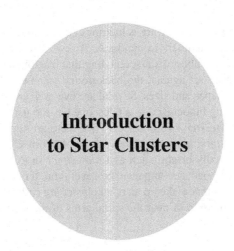

Introduction
to Star Clusters

Rationale Behind the Guide

Star clusters are among the most intriguing, amazing, and beautiful objects in the night sky. They can be young or old, large or small, bright or faint, and so on. But what is important, as they relate to this guide, is that seen in a telescope (or binoculars, or even the naked eye), they can be glorious, with a dazzling array of colors, brightnesses, and even shapes with arcs and streams, wisps of nebulosity, and dark dust lanes, making them literally breathtaking.

This guide was written with a specific type of observer in mind. This person tends to live in an urban or semi-urban environment, that is to say, light pollution will be an ever-present evil, but the observer may be able to go, from time to time, to a dark location and observe the night sky in all its glory. The person will most probably work during the day for a living, and so all-night observing sessions will be rare, with most observing taking place on weekends.

Bearing in mind the constraints of the preceding paragraph, the clusters in this guide were chosen so that nearly all observers can observe them with a variety of telescopes. These can range from small apertures (3–6 in.), moderate apertures (6–8 in.) and larger apertures[1] (8–10 in.), while some of the brighter clusters are binocular objects, and a few are naked eye. Many of the clusters will be familiar, but I have attempted to include some that are often ignored or passed over, or even, in a few instances, largely unknown to the amateur astronomy community.

Before I discuss the layout of the guide, observing techniques, and cluster classification, it is worth spending a short time discussing just what star clusters really are.

---

[1]The number of clusters that can be seen in very large telescopes is astounding. Literally hundreds more will be available. And if you live in the southern hemisphere, then the number of clusters is vast, due to the Large and Small Magellanic Clouds.

M. Inglis, *Observer's Guide to Star Clusters*, The Patrick Moore Practical Astronomy Series, DOI 10.1007/978-1-4614-7567-5_1, © Springer Science+Business Media New York 2013

Open Clusters

Open clusters, or *galactic clusters* as they are sometimes called, are collections of young stars containing anywhere from a dozen members to hundreds. A few of them (for example, *Messier 11* in Scutum) contain an impressive number of stars, equaling that of globular clusters, while others seem little more than a faint grouping set against the background star field. Such is the variety of open clusters that they come in all shapes and sizes. Several are over a degree in size, and their full impact can only be appreciated by using binoculars, as a telescope has too narrow a field of view. An example of such a large cluster is *Messier 44* in Cancer. Then there are tiny clusters, seemingly nothing more than compact multiple stars, as is the case with *IC 4996* in Cygnus. In some cases, all the members of the cluster are equally bright, such as *Caldwell 71* in Puppis, but there are others that consist of only a few bright members accompanied by several fainter companions, as is the case with *Messier 29* in Cygnus. The stars that make up an open cluster are called Population I stars, which are metal-rich and usually to be found in or near the spiral arms of the Milky Way Galaxy.

Size

The size of a cluster can vary from a few dozen light years across, as in the case of *NGC 255* in Cassiopeia, to about 70 light years across, as in either component of *Caldwell 14,* the Perseus Double Cluster.

Distribution in the Sky

An interesting aspect of open clusters is their distribution in the night sky. You may be forgiven in thinking that they are randomly distributed across the sky, but surveys show that although well over a thousand clusters have been discovered, only a few are observed to be at distances greater than 25° above or below the galactic equator. Some parts of the sky seem to be very rich in clusters – i.e., Cassiopeia and Puppis – due to the absence of dust lying along these lines of sight, allowing us to see across the spiral plane of our galaxy. Many of the clusters mentioned here actually lie in different spiral arms, so as you observe them, you are actually looking at different parts of the spiral structure of our own Galaxy.

Appearance

The reason for the varied and disparate appearances of open clusters is the circumstances of their births. It is the interstellar cloud that determines both the number and type of stars that are born within it. Factors such as the size, density, turbulence, temperature, and magnetic field all play a role as the deciding parameters in star birth. In the case of giant molecular clouds, or GMCs, the conditions can give rise to both O- and B-type giant stars[2] along with solar-type dwarf stars, whereas in small

---

[2]For a detailed explanation of the classification of stars, and even more information about their life cycles, see the references listed at the end of this book.

molecular clouds (SMCs) only solar-type stars will be formed, with none of the luminous B-type stars. An example of an SMC is the Taurus Dark Cloud, which lies just beyond the *Pleiades* cluster.

Stars are not born in isolation. Nor are they born simultaneously. Some clusters have bright young O and B main-sequence stars, while at the same time contain low-mass members, which may still be in the process of gravitational contraction (for example, the star cluster at the center of the Lagoon Nebula). In a few cases, the star production in a cluster is at a very early stage, with only a few stars visible, the majority still in the process of contraction and hidden within the interstellar cloud.

A perfect example of such a process is the open cluster within *Messier 42*, the Orion Nebula. The stars within the cluster, the *Trapezium*, are the brightest, youngest, and most massive stars in what will eventually become a large cluster containing many A-, F-, and G-type stars. However, the majority of those are blanketed by the dust and gas clouds and are only detectable by their infrared radiation.

Age

As time passes, the dust and gas surrounding a new cluster will be blown away by the radiation from the O-type stars, resulting in the cluster's becoming visible in its entirety, such as in the case of the young cluster *Caldwell 76* in Scorpius.

Once a cluster has formed, it will remain more or less unchanged for at least a few million years, but then changes within the cluster may occur. Two processes are responsible for changes within any given cluster. The evolution of open clusters depends on both the initial stellar content of the group and the ever-pervasive pull of gravity. If a cluster contains O-, B-, and A-type stars, these stars will eventually become supernovae, leaving the cluster with slower evolving, less massive, and less luminous members of type A and M stars. A famous example of such a cluster is *Caldwell 94, the Jewel Box* in Crux, which is a highlight of the southern sky, and, alas, unobservable to northern hemisphere observers. However, these too will become supernovae, with the result that the most luminous members of a cluster will, one by one, disappear over time. This doesn't necessarily mean the demise of a cluster, especially those clusters that have many tens or hundreds of members. But some, which consist of only a few bright stars, will seem to meld into the background star field. However, even those clusters that have survived the demise of their brighter members will eventually begin to feel the effect of a force that pervades everywhere – the Galaxy's gravitational field.

As time passes, the cluster will be affected by the influence of nearby globular clusters and the interstellar matter itself, as well as the tidal force of the Galaxy. The cumulative effect of all of these encounters will result in some of the less massive members of the cluster acquiring enough velocity to escape from the cluster. Thus, given enough time, a cluster will fade and disperse. (Take heart, as this isn't likely to happen in the near future so that you would notice. The *Hyades* star cluster, even after having lost most of its K- and M-type dwarf stars, is still with us after 600 million years!).

Color

Many experienced observers know that color in observed stars is best seen when contrasted with a companion(s). Thus, an open cluster presents a perfect opportunity for observing star colors. Many clusters, such as the ever and rightly popular *Pleiades,* are all a lovely steely blue color. On the other hand, *Caldwell 10* in Cassiopeia has contrasting bluish stars along with a nice orange star. Other clusters have a solitary yellowish or ruddy orange star along with fainter white ones, such as *Messier 6* in Scorpius. An often-striking characteristic of open clusters is the apparent chains of stars that are seen. Many clusters have stars that arc across apparently empty voids, as in *Messier 41* in Canis Major.

Globular Clusters

In the night sky are many compact and spherical collections of stars. These stars clusters are called *globular clusters*. These are metal-poor stars, often called Population II stars, and are usually to be found in a spherical distribution around the galactic center at a radius of about 200 light years. Furthermore, the number of globular clusters increases significantly the closer one gets to the galactic center. This means that particular constellations that are located in a direction towards the galactic bulge have a high concentration of globular clusters within them, such as Sagittarius and Scorpius.

The origin and evolution of a globular cluster is very different from that of an open or galactic cluster. All the stars in a globular cluster are very old, with the result that any star earlier than a G- or F-type star will have already left the main sequence and be moving toward the red giant stage of its life. In fact, new star formation no longer takes place within any globular clusters in our Galaxy, and they are believed to be our Galaxy's oldest structures. This means the youngest of the globular clusters is still far older than the oldest open cluster.

The origin of globular clusters is still hotly debated, with some researchers suggesting that they may have been formed within the proto-galaxy clouds that went to make up our Galaxy. However, in recent years two other origins for the globular clusters have come to light. Firstly, several seem to have been literally ripped from other nearby smaller galaxies, by the gravitational attraction of the Milky Way, and now orbit the core of our Galaxy, for example, *Palomer 12*. Secondly, it is known that our Galaxy has destroyed several smaller galaxies by a process called galactic cannibalism. The remnants of these devoured galaxies are, in some instances, believed to be some of the globular clusters we see, for example, *Omega Centauri*.

As previously mentioned, globular clusters are old; this means that all of the high-mass stars in a globular cluster evolved into red giants a long time ago. What remain are the low-mass main-sequence stars that are very slowly turning into red giants.

From an observational point of view, globular clusters can be a challenge. Many are only visible in optical instruments, from binoculars to telescopes, but a few are visible to the naked eye. There are about 200 globular clusters,[3] ranging in size from 60 to 150 light years in diameter. They all lie at vast distances from the Sun and are about 60,000 light years from the galactic plane. The nearest globular clusters (for example, *Caldwell 86* in Ara) lie at a distance of over 6,000 light years, and thus the clusters are difficult objects for small telescopes. That is not to say they can't be seen; rather, it means that any structure within the cluster will be difficult to observe. Even the brightest and biggest globular will, in most cases, require apertures of at least 15 cm for individual stars to be resolved. However, if large-aperture telescopes are used, these objects will be magnificent. Some globular clusters have dense concentrations towards their center, while others may appear as rather compact open clusters. In some cases, it is difficult to say where the globular cluster peters out and the background stars begin.

Stellar Associations and Streams

There exists another type of star grouping, which is much more ephemeral and spread over a very large region of the sky, and although not strictly associated with star clusters, they are very, very large groups of stars.[4] They are included in the guide because they are impressive objects, and what's more can be seen with the naked eye.

---

[3]Not all of these are visible in the range of amateur telescopes as mentioned in this book.

[4]These are large groups of stars, so, can be thought of as clusters – with only a small distortion of the truth.

A stellar association is a loosely bound group of very young stars. It may still be swathed in the dust and gas cloud within which the stars formed, and star formation may still be occurring within the cloud. Where they differ from open clusters is in their enormity, covering both a sizable angular area of the night sky and at the same time encompassing a comparably large volume in space. As an illustration of this huge size, the *Scorpius-Centaurus Association* is about 700 by 760 light years in extent, and it covers about 80°.

There are three types of stellar associations:

OB Associations. These contain very luminous O- and B-type main-sequence, giant, and super-giant stars.

B Associations. These contain only B-type main-sequence and giant stars but with an absence of O-type stars. These associations are just older versions of the OB association, and thus the faster-evolving O-type stars have been lost to the group as supernovae.

T Associations. These are groupings of T Tauri-type stars. They are irregular variable stars that are still contracting and evolving toward being A-, F-, and G-type main-sequence stars. More often than not they will be shrouded in dark dust clouds, and those that are visible will be embedded in small reflection and emission nebulae.

The lifetime of an association is comparatively short. The very luminous O-type stars are soon lost to the group as supernovae, and, as usual, the ever-pervasive gravitational effects of the Galaxy soon disrupt the association. As time passes, the B-type stars will disappear through stellar evolution, and the remaining A-type and later stars will now be spread over an enormous volume of space, the only common factor among them their motion through space. At that point, the association is called a stellar stream.

An example of such a stream and one that often surprises the amateur is the *Ursa Major Stream*. This is an enormous group of stars, with the five central stars of Ursa Major (the Plough) being its most concentrated and brightest members. The stream is also known as the *Sirius Supercluster* after its brightest member. The Sun actually lies within this.

Observing Star Clusters

I are now going to make a somewhat odd statement: "All clusters are not created equal." What I mean by this is that some clusters are spectacular, with a glittering display of colored doubles and triples, in curving loops, chains, and arcs, interspersed with dark starless voids and lanes. These clusters will literally make your jaw drop, and you will revisit them time and time again. Others, it has to be said, even with the largest apertures and with the highest magnifications, will be nothing more than a tiny, hazy spot. And once found and observed, you may well ask yourself why you bothered to observe it in the first place.

I regard all clusters worthy of observation, but honestly, in some cases, a few clusters will be very unimpressive. Nevertheless I have included both types, as the former are great to look at, while the latter can be used as an exercise in your observing techniques and a test of your telescope optics. Maybe it is best put this way: "All clusters are equal, but some clusters are more equal than others."

Clothing

This may seem a odd topic, but, as experienced observers know only too well, if you are not kept reasonably warm while you observe, then in only a very short time you will be far too cold to even make any pretence of observation.[5] Basically, common sense should prevail, and even if you don't actually wear any thick clothing to begin with, it should be available as the temperature drops. Nothing spoils a clear night's observing more than having cold hands and feet.

A last piece of advice is to know when it's time to stop observing! This is usually when your teeth are chattering, and your body is shaking from the cold.[6] This is the time to go indoors and have a hot drink, looking back in your mind over the incredible sights you have seen earlier that evening.

Recording Your Observations

This is a subject that, for some unknown reason, is often ignored by many amateur astronomers but is quite important. If you are just casually observing, sweeping the sky at leisure, then recording your observations may indeed be superfluous. But once you start to search out and observe specific objects, then note-taking should become second nature, if only to make a checklist of the items you have seen.

Basically, you should record the object viewed, the time of the observation, the telescope used (with eyepiece and magnification), the seeing and transparency, and maybe a brief description of the object and a sketch. If you wish these notes can then be copied up into more formal notes later, preferably the next day. (Your initial notes should always be made at the telescope, so as to keep as accurate a record as possible.)

You'll notice this book has left some room at the end of each constellation section for you to make brief notes on what you have observed. Some observers use a separate book or notepad for each type of object observed. A pocket dictation machine can be a quick and easy substitute for the notebook (apart from the sketch!). Whatever method you use, keeping a record will help you keep track of your observations and will help you to become a better observer.

Classification

Because open clusters display such a wealth of characteristics, different parameters are assigned to a cluster that describe its shape and content. For instance, a designation called the Trumpler type is often used. It is a three-part designation that describes the cluster's degree of concentration – that is, from a packed cluster to one that is evenly distributed, the range in brightness of the stars within the cluster, and finally the richness of the cluster, from poor (fewer than 50 stars) to rich (more than 100). The full classification is:

Trumpler Classification for Open Star Clusters

Concentration

I. Detached – strong concentration of stars towards the center.
II. Detached – weak concentration of stars towards the center.

---

[5] Those of us familiar with the climate of northern Europe and the UK know that even in early and late summer, when the days may be warm, and sometimes hot, the nights can still be quite chilly.

[6] Believe it or not, many observers myself included, have reached this stage on several occasions, not wanting to waste a minute of a clear night!

II. Detached – no concentration of stars towards the center.
IV. Poor detachment from background star field.

Range of Brightness

1. Small range
2. Moderate range
3. Large range

Richness of Cluster

p-Poor (fewer than 50 stars)
m-Moderate (50 to 100 stars)
r-Rich (more than 100 stars)
n-Cluster within nebulosity

Shapley-Sawyer Classification for Globular Star Clusters

As in the case of open clusters, there exists a classification system for globular clusters, the Shapley-Sawyer Concentration Class, where Class I globular clusters are the densest and Class XII the least dense. The ability of an observer to resolve the stars in a globular actually depends on how condensed the cluster is, and so the scheme will be used in the descriptions, but it is really only useful for those amateurs who have large-aperture instruments. Nevertheless, the observation of these clusters, which are among the oldest objects visible to amateurs, can provide you with breathtaking, almost three-dimensional, aspects.

Constellation Data and Cluster Lists

The constellations are presented alphabetically,[7,8] and so are at their best for observing at different times of the year. The following table will help you locate the clusters and constellation throughout the year.[9,10]

January
Cancer, Canis Minor, Carina, Gemini, Lynx, Monoceros, Puppis, Volans

February
Cancer, Carina, Pyxis, Vela

---

[7] You may notice that some constellations are absent from the list. That is because they do not have any clusters in them.

[8] The material was presented this way after much consultation with many astronomers.

[9] If a constellation appears in 2 months, it is because it culminates on either the last, or first, day of a month.

[10] Any cluster or constellation can of course be observed earlier or later than this date as it rises about 4 min earlier each night, nearly ½h each week, and thus about 2 h a month. To observe any cluster or constellation earlier than its culmination date, you will have to get up in (or stay up to) the early hours of the morning. To observe later than the culmination date will mean observing earlier in the evening, and vice versa.

March
Centaurus, Crux, Hydra, Musca, Ursa Major

April
Canes Venatici, Circinus, Coma Berenices, Virgo

May
Apus, Boötes, Libra, Lupus, Norma, Serpens, Triangulum Australe, Ursa Minor

June
Ara, Corona Austrina, Hercules, Ophiuchus, Scorpius, Scutum

July
Corona Austrina, Cygnus, Lyra, Pavo, Sagitta, Sagittarius, Scutum, Telescopium, Vulpecula

August
Aquarius, Capricornus, Delphinus, Lacerta, Octans, Pegasus

September
Aquila, Cepheus, Pegasus, Sculptor, Tucana

October
Andromeda, Cassiopeia, Hydrus, Triangulum

November
Fornax, Horologium, Perseus, Taurus

December
Auriga, Camelopardalis, Columba, Dorado, Lepus, Mensa, Orion, Pictor, Taurus

In the constellation descriptions, I have used the following nomenclature:

- The name of the constellation
- The abbreviated name of the constellation
- The genitive name of the constellation (Every constellation name has two forms: the *nominative*, for use when one is talking about the constellation itself, and the *genitive*, or possessive, which is used in star names). For an example, Capella, the brightest star in the constellation Auriga (nominative form), is also called Alpha Aurigae (genitive form), meaning literally "the Alpha of Auriga."
- The common translation for the constellation
- The (approximate) latitude range within which the constellation can be seen
- The month during which the constellation will be at its highest point in the sky at midnight, known as its culmination,[11] or transit

Regarding specific clusters, the following information is given:

First line:

- Official reference name, i.e., NGC 5272, IC 4321
- Popular name, or another reference name, i.e., Collinder 42, Messier 3, Caldwell 10, Herschel 61

---

[11] It can of course be seen earlier or later in the month given, but of course it may not be at its highest point in the sky at midnight.

- Position, in right ascension, and declination. RA, Dec
- Type of object
  - OC (open cluster)
  - GC (globular cluster)
  - Ast (asterism)
  - Neb (nebula)
  - ?-unknown

Second line:

- Visual magnitude,[12] v, the combined magnitude of all stars in cluster, or photographic magnitude, p.
- Object size[13] in arc minutes ($\oplus$)
- Approximate number of stars in the cluster, but bear in mind that the number of stars seen will depend on magnification and aperture, and will increase when large apertures are used. Thus the number quoted is an estimate using modest aperture.
- Trumpler designation, or Shapley-Sawyer concentration class
- Level of difficulty on observing the cluster based on the magnitude, size, and observing conditions, plus ease of finding the cluster.

Constellation Maps

Each constellation has its own star map to help you to locate the clusters within them. But note here and now that these maps are *very simple* and are only meant to be signposts to their approximate location of the objects in the sky. They are not meant to be detailed charts, and should not be treated as such, but rather as pointers to the approximate locations of the objects under discussion.

There exist a plethora of planetarium software programs and apps for use on smart phones, iPads or the equivalent and desktop computers. Several of them are free, and nearly all are excellent. These can be used for making detailed star maps for use at the telescope, if needed.

Also, in recent years there has been in revolution in observational astronomy, specifically with the introduction of go-to telescopes. These have truly revolutionized astronomy and can, in some instances, make star atlases obsolete and the locating of very faint objects a trivial exercise.

---

[12] The quoted magnitude of a cluster is the integrated magnitude. This is the combined visual magnitude of all its components, as if the cluster's light was combined into a star like point. It may be the result of only a few bright stars, or, on the other hand, may be the result of a large number of faint stars. But beware! The cluster is not a point of light but spread across a discernible diameter, so, as the light spreads out, its intensity will drop off rapidly, causing the cluster to become fainter. Thus, the quoted magnitude may appear considerably dimmer than a star of the same absolute magnitude. Different observers will see differing magnitudes. Treat the given value with a certain amount of caution.

[13] Also, the diameter of a cluster is often misleading, as in most cases it has been calculated from photographic plates, which, as experienced amateurs will know, can bear little resemblance to what is seen at the eyepiece.

However, even with a computerized telescope, it is still vitally important to obtain a good star atlas. There are many now in print, and so choose the one that best suits your needs.

Here, now, is a list of constellations found in this guide:

Constellation List

Andromeda	Delphinus	Perseus
Apus	Dorado	Pictor
Aquarius	Fornax	Puppis
Aquila	Gemini	Pyxis
Ara	Hercules	Sagitta
Auriga	Horologium	Sagittarius
Boötes	Hydra	Scorpius
Camelopardalis	Hydrus	Sculptor
Cancer	Lacerta	Scutum
Canes Venatici	Lepus	Serpens
Canis Major	Libra	Taurus
Canis Minor	Lupus	Telescopium
Capricornus	Lynx	Triangulum
Carina	Lyra	Triangulum Australe
Cassiopeia	Mensa	Tucana
Centaurus	Monoceros	Ursa Major
Cepheus	Musca	Ursa Minor
Circinus	Norma	Vela
Columba	Octans	Virgo
Coma Berenices	Ophiuchus	Volans
Corona Austrina	Orion	Vulpecula
Crux	Pavo	
Cygnus	Pegasus	

How to Use This Guide

You may think this penultimate section has an odd title, but bear with it.

There exist today many wonderful books that describe in great detail star clusters, with photographs of each object along with detailed star maps, drawings of each object under discussion, and a history of the object, etc. Such books are truly wonderful, printed on the best quality paper, and a must purchase for any observer's library.[14] However, you would never dream of writing in them, or taking them out to the telescope in the evening. They are more for reference in planning an observing session.

This book is different. This guide of star clusters is to be used! Take it out at night when you observe, write in it, comment on what you see, draw in it. Use it as you will, as it is not meant to be left on your bookshelf but rather used at the telescope as you seek out these amazing objects.

And Finally…

Having read this chapter, you should now have some idea of what's in store for you, but before you rush outside, keep these thoughts in mind. Astronomy, like any other hobby or pastime (or lifelong devotion!), gets easier with time. The longer you spend observing, the better you will become at it. But remember that success in seeing all the objects that you observe – or try to observe – will depend on many things: the seeing, the time of year, the instruments used, and even your state of health! Don't be despondent if you can't find an object on your first attempt.[15] It may be beyond your capabilities, at that particular time, to see it. Just record the fact of your non-observation, and move onto something new. You'll be able to go back to the elusive object another day or month. It will still be there. You may see differing star colors to those mentioned here, or different asterisms. This doesn't matter, as nobody's eyes are identical, and neither are their observing locations or sky conditions. Also, take your time; you don't need to rush through the observations. Try spending a long time on each object you observe. In the case of several clusters, it can be very instructive, fascinating, and often breathtaking to let the cluster drift into the telescope's field of view. You'll be surprised at how much more detail you will seem to notice.

Although I have tried to include all the famous objects, if any of your favorites have been omitted, then I apologize! To include everybody's favorites would be a nice idea, but an impossible task.[16]

Now get started! The universe awaits you!

---

[14] They also quite heavy, and so could probably be used as an emergency counterweight.

[15] It is often useful to be able to determine the night sky's observing conditions (light pollution, haze, cloud cover, transparency) before starting an observing session, so as to determine what types of objects will be visible and even allow you to decide whether observing is viable at all. A good way to do this is to use a familiar constellation, which should be observable every night of the year, and estimate what stars in the constellation are visible. If only the brighter stars are visible, then this would limit you to only bright stellar objects, while if the fainter stars in the constellation can be seen, then conditions may be ideal to seek out the more elusive, and fainter objects. A favorite constellation used by many amateurs for just such a technique is Ursa Minor, the Little Bear. If, once outside, you can see υ UMi (magnitude 5.2) from an urban site, then the night is ideal for deep-sky observing. However, if υ UMi is not visible, then the sky conditions are not favorable for any serious deep-sky observing, but casual constellation observing may be possible. If the stars δ, ε, and ζ UMi, located in the "handle" of the Little Bear, are not visible (magnitudes 4.3, 4.2 and 4.3, respectively) then do not bother observing at all, but go back indoors and peruse this book.

[16] Compiling such an guide that includes a terrific amount of data was a difficult, but tremendously rewarding task. However, there may be times when I have inadvertently made a typographic error, such as in the size of a cluster, or its magnitude, etc. If this occurs then I apologise, as any such errors are mine and mine alone.

The Constellations and Their Star Clusters

M. Inglis, *Observer's Guide to Star Clusters*, The Patrick Moore Practical Astronomy Series, DOI 10.1007/978-1-4614-7567-5_2, © Springer Science+Business Media New York 2013

Andromeda

Fast Facts

Abbreviation: And	Genitive: Andromedae	Translation: Princess of Ethiopia
Visible between latitudes 90° and −40°		Culmination: October

Star Clusters

Herschel 69	NGC 7686	23ʰ 30.1ᵐ	+49° 08′	OC
5.6 m	⊕ 15′	80	IV 1 p	Moderate

A sparse and widely dispersed cluster with many 10th and 11th magnitude stars, the cluster can be glimpsed in binoculars as a diffuse glow centered on a yellowish 6.5 magnitude star (it has also been called reddish-orange!). As larger apertures are used, fainter members appear and stream off in several chains. It lies at a distance of 3,000 light years and is around 13 light years in diameter. It was discovered in 1787.

Collinder 33	NGC 752	01ʰ 57.8ᵐ	+37° 41′	OC
5.7 m	⊕ 45′	77	III 1 m	Easy

Best seen in binoculars, or even at low powers in a telescope, this is a large, loosely structured group of stars containing many chains and double stars. It l lies about 5° south-southwest of γ Andromedae. Often underrated by observing guides, it is worth seeking out. It is a cluster of intermediate age.

Notes

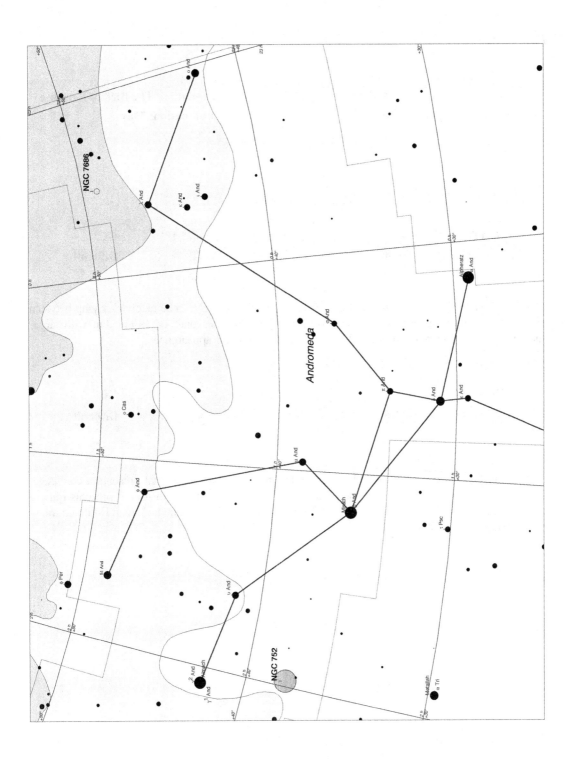

Apus

Fast Facts

Abbreviation: Aps	**Genitive: Apodis**	**Translation: The Bird of Paradise**
Visible between latitudes 5° and −90°		**Culmination: May**

Clusters

IC 4499	–	15ʰ 00.3ᵐ	−82° 13′	GC
10.1 m	⊕ 8′	–	XI	Difficult

This is a globular cluster that needs a large aperture in order for it to be resolved. Lying between π^1 and π^2 Apodis it appears as an unresolved soft glow in telescopes of 30 cm, but with larger apertures of, say, 35 cm and up, it will begin to show a granular appearance.

NGC 6101	–	16ʰ 25.8ᵐ	−72° 12′	GC
9.2 m	⊕ 5′	–	X	Difficult

As with the previous entry, IC 4499, this is another globular cluster that presents a challenge. Larger apertures are needed to see and resolve the cluster, but it is worth it as it presents quite a spectacle and takes high magnification well. With an aperture of 40 cm and larger, and still and transparent skies, the core shines out!

Notes

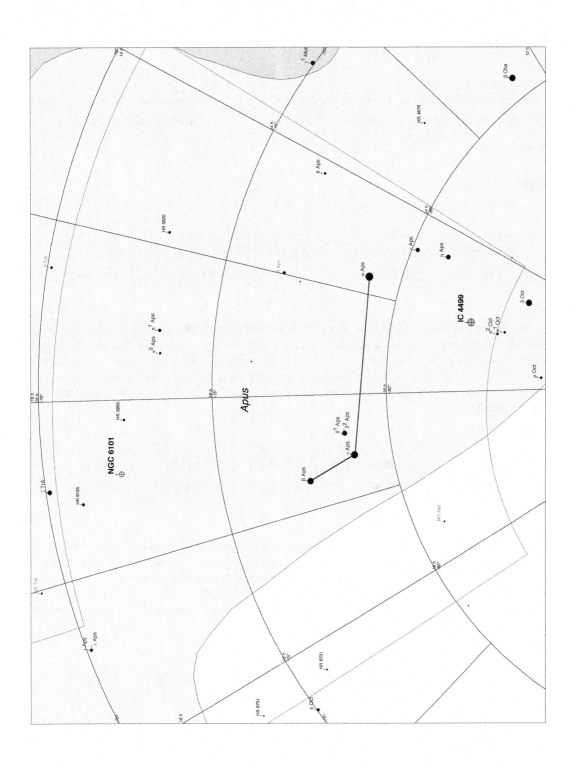

Aquarius

Fast Facts

Abbreviation: Aqr	Genitive: Aquarii	Translation: The Water Bearer
Visible between latitudes 65° and −90°		Culmination: August

Clusters

NGC 6981	Messier 72	20ʰ 53.5ᵐ	−12° 32′	GC
9.3 m	⊕ 5.9′		IX	Moderate

At a distance of about 53,000 light years, this is faintest globular cluster cataloged by Messier. In binoculars it will appear as just a tiny, hazy point of light, but in telescopes or aperture 20 cm and larger its true nature becomes apparent. The use of averted vision may help you to see any detail within the cluster. Lying in the halo of the galaxy, and rotating in a retrograde direction, that is, opposite to that of the Sun, has led some astronomers to speculate it originated in another galaxy that was cannibalized by ours!

NGC 6994	Collinder 426	20ʰ 59.0ᵐ	−12° 38′	OC/Ast
8.9 m	⊕ 2.8′	4	IV 1 p	Easy

Also known as Messier 73, this is something of an enigma; perhaps it shouldn't really be classified as an open cluster, as it consists of only four stars! Cataloged possibly when Messier was having a bad day, it is still nice. A small grouping of stars like this is often called an asterism. There is considerable debate as to whether the stars are related to each other, or are the result of a chance alignment. One line of research suggests M73 is a "possible open cluster remnant (POCR)," whereas later work suggests otherwise.

NGC 7089	Messier 2	21ʰ 33.5ᵐ	−00° 49′	GC
6.5 m	⊕ 12.9′	II		Easy

This is a very impressive globular cluster and is often referred to as the showpiece of the constellation. It can be seen with the naked eye, although averted vision will be necessary. However, as it is located in a barren area of the sky it can prove difficult to locate. But when found it is a rewarding object, and even in large binoculars its oval shape is apparent. Telescopes will show its bright core, and larger instruments will show several star chains snaking out from the core. Believed to be about 40,000 light years away, it contains over 100,000 stars and takes over a billion years to orbit the Milky Way.

NGC 7492	Herschel V-558	23^h 08.4^m	−15° 37′	GC
11.4 m	⊕ 6.2′		XII	Difficult

This globular cluster is extremely faint, and even in apertures of 25 cm will be just visible with averted vision. With telescopes of 40 cm, it will still be unresolved, but at least a diffuse glow will be seen.

Notes

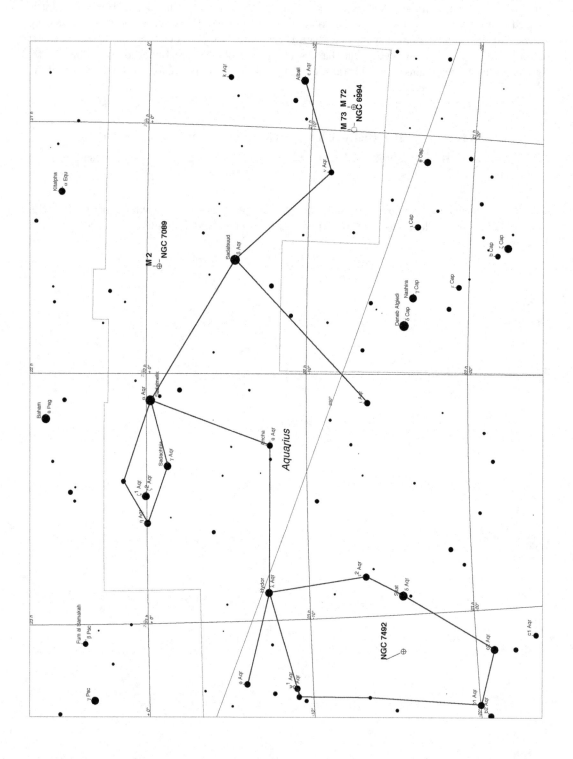

Notes (cont.)

Aquila

Fast Facts

Abbreviation: Aql	Genitive: Aquilae	Translation: The Eagle
Visible between latitudes 85° and −75°		Culmination: September

Clusters

NGC 6709	Collinder 392	18ʰ 51.5ᵐ	+10° 21′	OC
6.7 m	⊕ 13′	45	III 2 m	Moderate

A difficult object to locate for binoculars, as it will be unresolvable and present itself as a very faint blob. Nevertheless this presents a good challenge for you to hone your observing skills. For small telescopes it presents a nice rich cluster easily resolved. Larger apertures show fainter stars with a nice pair of yellow and blue stars on the western edge. As Aquila lies in the vicinity of the Great Rift, an enormous cloud of dust and gas, open clusters like this are few and far between, and the center of the cluster itself contains many dark and empty patches.

NGC 6738	–	19ʰ 01.4ᵐ	+11° 36′	OC
8.3 m	⊕ 15′	75	IV 2 p	Moderate

Another very faint open cluster that will need a larger aperture in order for any detail to be seen. Consisting of mainly 9.5 magnitude and fainter stars, it appears as a large and irregular cluster with perhaps as many as 75 stars.

NGC 6749	–	19ʰ 05.1ᵐ	+01° 47′	GC
12.4 m	⊕ 6.3′		–	Moderate

Observers with telescopes of aperture 25 cm and smaller move on! This globular cluster needs at least at least a 30 cm telescope, and even then it will only appear as faint and unresolved. Nevertheless it is located in a rich area of the Milky Way, so finding it would be a worthwhile challenge.

NGC 6755	–	19ʰ 07.8ᵐ	+04° 16′	OC
7.5 m	⊕ 15′	100	II 2 r	Moderate

An easily found open cluster in small telescopes, standing out from the background star field. This is a nice object, as increasing the aperture and magnification will show many faint stars, dark lanes, and star chains. The open cluster Czernik 39 lies about 7′ NNW.

NGC 6756	–	19ʰ 08.7ᵐ	+04° 42′	OC
10.6 m	⊕ 4′	100	I 1 m	Moderate

This is a much fainter and smaller open cluster than its neighbor, NGC 6755. With a 20 cm telescope, maybe a dozen stars can be seen against a vague, unresolved background. Larger apertures will just show even fainter stars.

NGC 6760	–	19ʰ 11.2ᵐ	+01° 02′	GC
9.1 m	⊕ 6.6′		IX	Moderate

This is a faint symmetrical globular cluster with a just perceptible brighter core. High-power binoculars should be able to locate this cluster, and even with a medium aperture telescope of aperture 16 cm it should present no problems. Knowledge of the use of setting circles would be useful, as would a computer-controlled telescope.

Palomer 11	–	19ʰ 45.14ᵐ	–08° 0.5′	GC
9.8 m	⊕ 10′	75	XI	Difficult

This is a very faint globular cluster that has proved difficult to see. In fact, several observers have expressed doubt that they ever saw it, even under dark skies and using averted vision. Those lucky enough to have seen it report it as extremely faint, with no brightening toward the center. This is a definite challenge!

NGC 6773	–	19ʰ 15.0ᵐ	+04° 53′	
–m	⊕ 12′ × 7′			Difficult
NGC 6775	–	19ʰ 16.8ᵐ	–00° 55′	
–m	⊕ 15′			Difficult
NGC 6795	–	19ʰ 26.0ᵐ	+03° 31′	
–m	⊕ 30′ × 15′			Difficult
NGC 6828	–	19ʰ 50.4ᵐ	+07° 55′	–
11.5 m	⊕ 3′			Difficult

The above four clusters are included as an observing exercise only. They are often classified as "nonexistent" in the *RNGC*,[1] as they are very difficult to distinguish from the background star fields. Nevertheless, they will pose a challenge to those observers who are lucky enough to have large aperture telescopes of at least 30 cm.

Notes

---

[1] The Revised New General Catalogue.

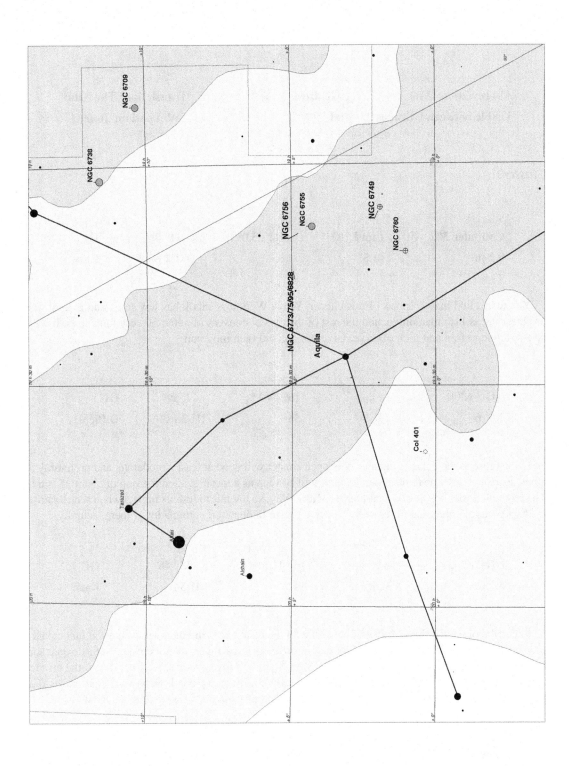

Fast Facts

Abbreviation: Ara	Genitive: Arae	Translation: The Altar
Visible between latitudes 25° and –90°		Culmination: June

Clusters

Collinder 307	Lund 709	16ʰ 35.0ᵐ	–51° 00′	OC
9.2 m	⊕ 5′	50	III 2 p	Easy

You won't find many reports of this cluster. Why? Well, it is small, has few stars, and is faint. Its brightest star is 11th magnitude, and the rest of the cluster consists of about 50 very faint stars. It will be a struggle to find and then observe. Give it a go, and then move on.

NGC 6200	–	16ʰ 44.1ᵐ	–47° 28′	OC
7.4 m	⊕ 12′	35	III 2 m	Difficult

Lying close to the galactic equator, this open cluster will need at least a moderate, and preferably a large, aperture telescope to be appreciated. It will be seen as a loose, scattered group of about 35 stars. The problem is that it is in a rich part of the Milky Way, so it tends to lose its impact when seen against the background. Larger apertures will of course reveal fainter stars, quite a lot of them, actually.

NGC 6193	–	16ʰ 41.3ᵐ	–48° 46′	OC
5.2 m	⊕ 15′	14	II 3 p	Easy

Reports abound that this cluster can be seen with the naked eye, and thus will be easy in binoculars. In a small telescope, it seems more of a random group of stars rather than a cluster, but as expected, increasing aperture will reveal its true nature. More stars will reveal themselves, but due to the erratic aspect of the cluster, the member stars tend to meld with the background. However, what makes the cluster far more impressive is that it is embedded in the emission and reflection nebula NGC 6188.

Hogg 22	–	16ʰ 46.6ᵐ	–47° 05′	OC
6.7 m	⊕ 1.2′		IV 3 p	Difficult

This cluster is a small group of about 10 stars, so in itself isn't much to write home about, but it makes a delightful companion to the next cluster in our list, NGC 6024.

NGC 6204	–	16h 46.6 m	–47° 01′	OC
8.2 m	⊕ 6′	45	I s m	Moderate

When both this cluster and the preceding one, Hogg 22, are seen together it makes a stunning sight. NGC 6204 consists of about four dozen 9th magnitude and fainter stars in quite a small area, so admittedly you will need a large aperture in order to appreciate this. But it is worth seeking out as double clusters as spectacular as this one, are few and far between.

NGC 6208	–	16h 49.4m	–53° 42′	OC
7.2 m	⊕ 18′		III 2 r	Difficult

This is an attractive cluster, with two nicely colored stars, orange and yellow, at its center. However, with small aperture it will hardly appear as a cluster at all, but more of slight enhancement of the background field. It only becomes apparent when seen in larger apertures and even then will need high magnification.

NGC 6250	–	16h 57.9m	–45° 56′	OC
5.9 m	⊕ 10′		IV 3 p	Easy

Easily glimpsed in a finder as a tiny group of barely, if at all, resolvable stars, this cluster is a nice object. Small telescopes will reveal more of its fainter magnitudes that cover a wide range of brightnesses, 11th–13th magnitudes.

IC 4651	Melotte 169	17h 28.8m	–49° 56′	OC
6.9 m	⊕ 10′	100	II 2 r	Moderate

This cluster is will appear as a collection of small groups and short arcs of stars. Maybe 100 can be seen with a large aperture and medium magnification.

NGC 6362	–	17h 32.0m	–67° 03′	GC
8.3 m	⊕ 15′		X	Easy

This globular cluster can be glimpsed in a finder and in binoculars as a small, fuzzy blob. To resolve any stars you will need at least a 15 cm telescope. But this cluster really comes into its own when a larger aperture as used, along with a high magnification. Then numerous stars will be seen set against a grainy seeming background glow.

NGC 6397	Dunlop 366	17^h 40.7^m	−53° 40′	GC
5.3 m	⊕ 32′		IX	Easy

When you observe this globular cluster, you are in fact looking at the second closest to us, the other being Messier 4. It is also one of the oldest, at nearly 13 billion years. Observationally, it is a gem. One of the sky's finest, although many observers have not seen it as it lies too far south for European and American telescopes. It can be seen easily in small telescopes, but the bigger the aperture (and magnification), the better. It has a bright halo with arcs of 11th and 12th magnitude stars apparently streaming from it. With a high magnification, the halo can fill the entire field. Do not miss this one if you can.

Notes

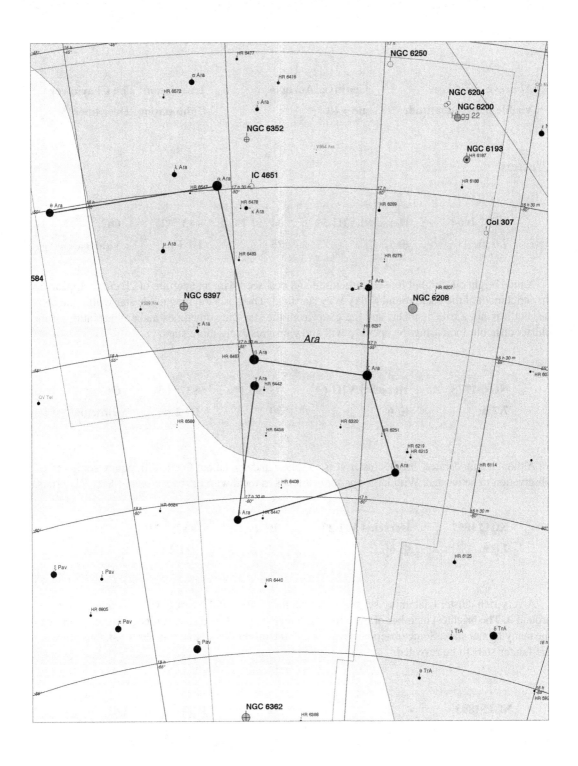

Auriga

Fast Facts

Abbreviation: Aur	**Genitive: Aurigae**	**Translation: The Charioteer**
Visible between latitudes 90° and −40°		**Culmination: December**

Clusters

NGC 1664	Herschel VIII-59	04ʰ 51.1ᵐ	+43° 42′	OC
7.6 m	⊕ 13′	25	III 1 p	Moderate

A nice bright cluster, but loosely structured and best seen with an aperture of 20 cm. It appears as an enrichment of the background Milky Way star field. There is a 7th magnitude star within the cluster, but it is not a true member, and the glare from the star can sometimes make observation of the cluster difficult. Increasing the aperture will show progressively more stars.

NGC 1778	Herschel VIII-61	05ʰ 08.1ᵐ	+37° 03′	OC
7.7 m	⊕ 6′	30	III 2 p	Difficult

Although a moderately bright cluster, it is so sparse and spread out that it will require some careful observation to be located. With large apertures it tends to meld with the background Milky Way stars.

NGC 1857	Herschel VII-33	05ʰ 19.1ᵐ	+39° 21′	OC
7.0 m	⊕ 6′	35	II 2 m	Easy

A very rich cluster containing several small chains of stars with starless voids located within and around it. The brightest member of the cluster is a nice orange-tinted star, but its glare can overpower the many fainter stars. Some observers try to occult the bright star so that it is obscured, thus allowing the fainter stars to be revealed.

NGC 1893	–	5ʰ 22.7ᵐ	33° 24′	OC
7.5 m	⊕ 12′	30	III 2 m n	Moderate

This is a faint cluster set among the stars of the Milky Way, thus making it appear difficult to make out. It will need telescopes of large aperture, say 20 cm and higher, for it to be really appreciated. Some nebulosity can be glimpsed with the use of a nebula filter. But be warned, in order to see this you will need a telescope of 30 cm or more and dark skies. Nevertheless, a decent observing challenge from urban locations.

NGC 1912	Collinder 67	05^h 28.7^m	+35° 50′	OC
6.4 m	⊕ 21′	75	III 2 m	Easy

Also known as Messier 38, this is one of the three Messier clusters in Auriga and is visible to the naked eye. It contains many A-type main-sequence and G-type giant stars, with a G0 giant being the brightest, magnitude 7.9. It's elongated in shape with several double stars and voids within it. Seen as a small glow in binoculars, it is truly lovely in small telescopes. It is an old galactic cluster with a star density calculated to be about eight stars per cubic parsec. There have been suggestions that this cluster, and Messier 36, may have been formed out of the same nebula. Only time and more research will tell.

NGC 1960	Collinder 71	05^h 36.3^m	+34° 08′	OC
6.0 m	⊕ 12′	70	II 3 m	Easy

Also known as Messier 36, this is about half the size of Messier 38, seen as a glow in binoculars. It is a large, bright cluster. Measurements indicate that it is 10 times farther away than the Pleiades. It contains a nice double star at its center. Owing to the faintness of its outlying members it is difficult to ascertain where the cluster ends. The cluster is visible to the naked eye. This is what British astronomer Dave Eagle had to say about the cluster: "A gorgeous sprinkling of stars with contrasting colors, much more concentrated towards its heart. But there is an added bonus; in a wide-angle view there is, close by, an asterism of stars that looks very much like a face smiling back at you. You just have to smile back at it."

Basel 4	–	05^h 48.5^m	+30° 13′	OC
9.1 m	⊕ 6′	15	II 1 p	Difficult

For those observers that like a real observing challenge, this is for you. A very faint and small cluster with about 15 stars, this will need an aperture of at least 30 cm, in order to be glimpsed, and an even larger aperture for any detail to be made out. It lies about 15 arc sec south of a 7.5 m star. Not a brilliant cluster by any means, but one that will test your observing skills.

NGC 2099	Collinder 75	05^h 52.3^m	+32° 33′	OC
5.6 m	⊕ 24′	150	II 1 r	Easy

Also known as Messier 37, this is another naked-eye cluster, and, in a word – superb! The finest cluster in Auriga. It really can be likened to a sprinkling of stardust, and some observers liken it to a scattering of gold dust. Contains many A-type stars and several red giants. It's visible at all apertures appearing as a soft glow with a few stars in binoculars, to a fine star-studded field in medium-aperture telescopes. In small telescopes using a low magnification it can appear as a globular cluster. The central star is colored a lovely deep red, although several observers report it as a much paler red, which may indicate that it is a variable star.

NGC 2126	Herschel VIII-68	06^h 02.5^m	+49° 52′	OC
10.2 m	⊕ 6′	40	II 1 p	Difficult

This cluster has been described as diamond dust scattered onto black velvet. This is a very faint but nice cluster, although it can prove a challenge to find. Oddly, for an open cluster, it lies about 1,000 light-years above the galactic plane.

NGC 2281	–	06^h 49.3^m	+41° 05′	OC
5.4 m	⊕ 15′	30	I 3 p	Moderate

Observers often miss this nice cluster, probably because it does not contain many stars. Nevertheless, it is worth trying to see this elusive object, as, at high magnifications and with large aperture (>25 cm), a nice asterism and a blue and red double star can be glimpsed. Studies show that the cluster has over 100 members and is about 10 light years in diameter.

Notes

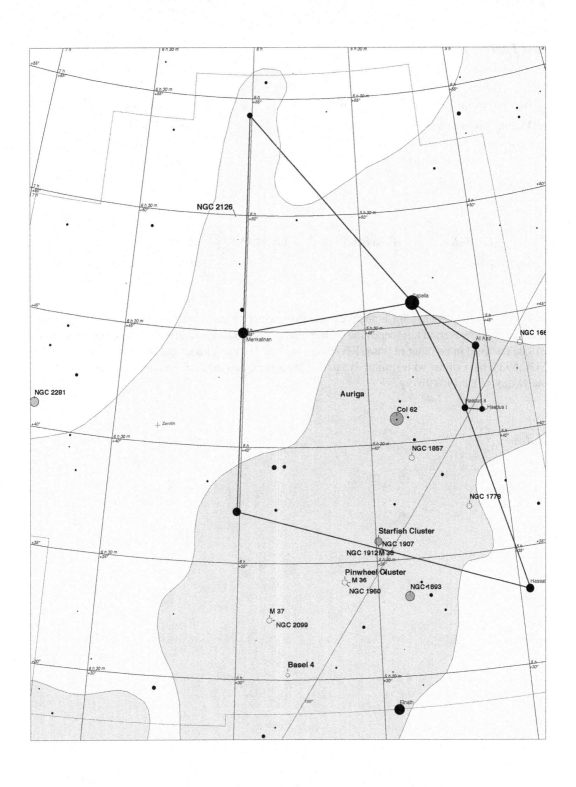

Boötes

Fast Facts

Abbreviation: Boo	Genitive: Boötis	Translation: The Bear Driver
Visible between latitudes 90° and −50°		Culmination: May

Clusters

NGC 5466	Herschel VI- 9	14ʰ 05.5ᵐ	+28° 32′	GC
9.0 m	⊕ 9.2′		XII	Moderate

This will be a challenge for binoculars, and even when located it will appear as a faint, hazy, and small glow. However, in telescopes the cluster has a resolvable core, and in large aperture many stars will be resolved in its outer regions. Research suggests that this cluster and another globular cluster, NGC 6934, have orbits with similar dynamical properties, and may be remnants of a disrupted satellite galaxy of the Milky Way.

Notes

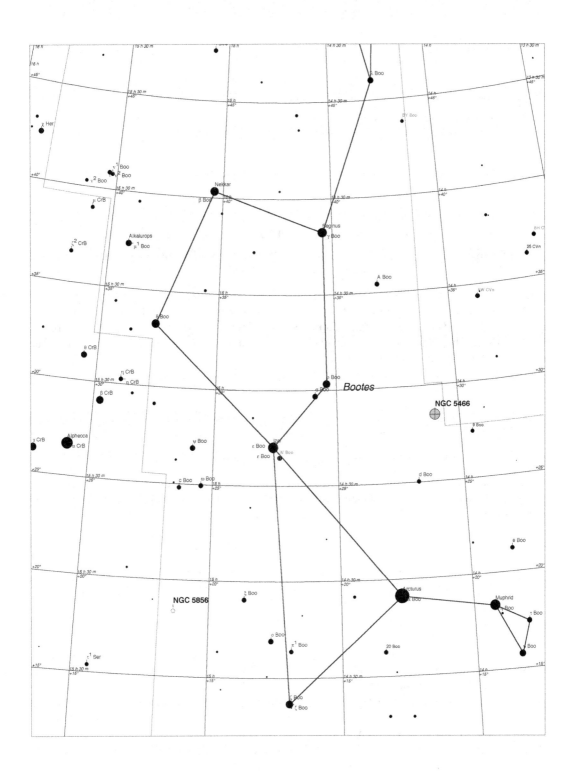

Camelopardalis

Fast Facts

Abbreviation: Cam	**Genitive: Camelopardalis**	**Translation: The Giraffe**
Visible between latitudes 90° and −10°		**Culmination: December**

Clusters

Stock 23	–	03ʰ 16.3ᵐ	+60° 02′	OC
5.6 m	⊕ 18′	25	III 3 p n	Easy

This is a little known cluster on the border of Camelopardalis-Cassiopeia. Binoculars will reveal several stars, but it is best viewed in medium-aperture telescopes that will show nicely a couple of yellow stars. It is bright and large but spread out. Recently it has been called Pazmino's Cluster after an observer, John Pazmino, championed its merits.

Tombaugh 5	–	03ʰ 47.8ᵐ	+59° 04′	OC
8.4 m	⊕ 17′	60	III 2 r	Easy

This is an open cluster often overlooked by observers even though it contains a lot of stars. The problem is the stars are all faint, 12th and 13th magnitude and fainter. Nevertheless in large apertures it does become apparent against the background star field.

NGC 1502	Herschel VII-47	04ʰ 07.8ᵐ	+62° 20′	OC
6.9 m	⊕ 8′	63	II 3 p	Moderate

This is a nice cluster and is easy to see, but can prove difficult to locate, even though it is in a relatively sparse area of the sky. Reports suggest that it visible to the naked-eye on clear nights. It is a rich and bright cluster but small and may resemble a fan shape, although this does depend on what the observer sees. What do you see? Also contained in the cluster are two multiple stars: Struve 484 and 485. The former is a nice triple system, but the latter is a true spectacle with nine components! Seven of these are visible in a telescope of 10 cm aperture, ranging between 7th and 13th magnitude. The remaining two components, 13.6 and 14.1 magnitude, should be visible in a 20 cm telescope. In addition, the system's brightest component, SZ Camelopardalis, is an eclipsing variable star, which changes magnitude by 0.3 over 2.7 days. What makes this cluster so special is its proximity to the asterism called Kemble's Cascade. This is a long string (2.5°) of 8th magnitude stars to the northwest of Herschel 47. The cascade is best seen in low-power binoculars. Overall, this is a worthwhile object to observe, lying as it does in a somewhat often-neglected part of the sky.

Collinder 464	–	05ʰ 22.0ᵐ	+73° 17′	OC
4.2 m	⊕ 120′	50	III 3 m	Easy

A large, very rich, irregular open cluster, with the distinction that it is best seen in binoculars, as viewing it in a telescope will dissipate the cluster. Contains many 5th, 6th, and 7th magnitude stars.

Notes

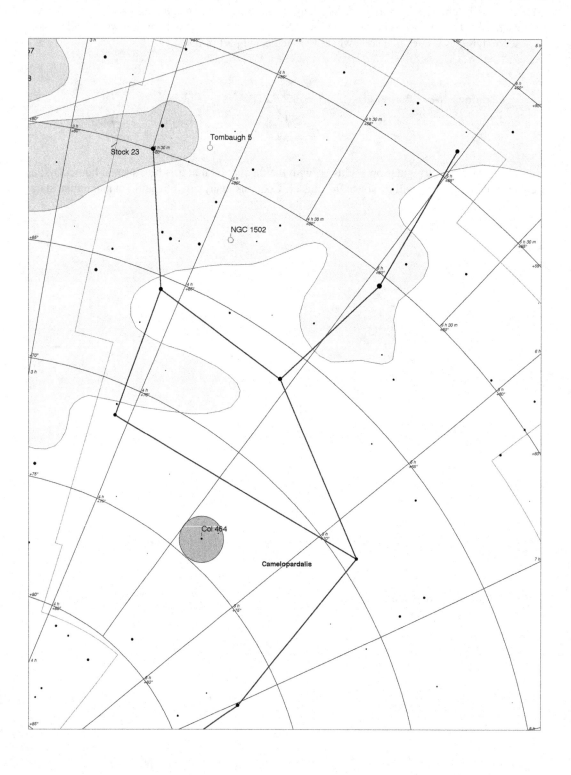

Notes (cont.)

Cancer

Fast Facts

Abbreviation: Cnc	Genitive: Cancri	Translation: The Crab
Visible between latitudes 90° and −60°		Culmination: January/February

Clusters

NGC 2632	Collinder 189	08ʰ 40.1ᵐ	+19° 40′	OC
3.1 m	⊕ 95′	60	II 2 m	Easy

Also known as Messier 44, this is a famous open cluster, called Praesepe (the Manger) or the Beehive. One of the largest and brightest open clusters from the viewpoint of an observer, as it can be seen in all sizes of aperture and with the naked eye. It's an old cluster, of about 700 million years, with a distance of 600 light years, having the same space motion and velocity as another famous cluster, the Hyades, that suggests a common origin for the two clusters. A nice triple star, Burnham 584, is located within Messier 44, located just south of the cluster's center. A unique Messier object in that it is brighter than the stars of the constellation within which it resides. Owing to its large angular size in the sky, about 1.6°, it is best seen through binoculars or a low-power eyepiece.

NGC 2682	Collinder 204	08ʰ 50.8ᵐ	+11° 49′	OC
6.9 m	⊕ 30′	200	II 2 m	Difficult

Also known as Messier 67, this open cluster is often overlooked owing to its proximity to Messier 44; nevertheless it is very pleasing. There is one caveat, however. The stars it is composed of are faint ones, so in binoculars it will be unresolved and seen as a faint misty glow. At a distance of 3,000 light years, it is believed to be very old, with recent research indicating 4 billion years, and has had time to move from the galactic plane, the usual abode of open clusters, to a distance of about 1,600 light years off the plane.

Notes

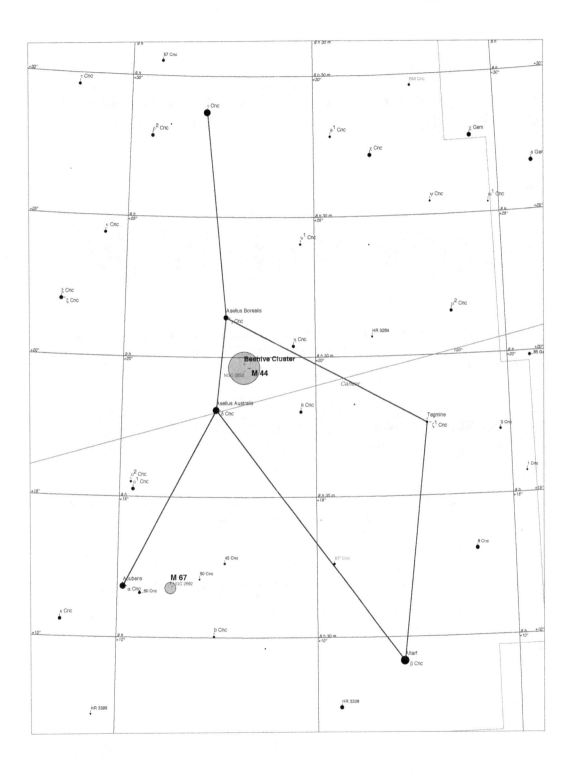

Canes Venatici

Fast Facts

Abbreviation: CVn	Genitive: Canum Venaticorum	Translation: The Hunting Dogs
Visible between latitudes 90° and −40°		Culmination: April

Clusters

Upgren 1	–	12^h 35.0^m	+36° 18′	OC/Ast
6.6 m	⊕ 14′	10	IV 2 p	Easy

Almost unknown to observers, this is a fairly inconspicuous cluster. Binoculars show about 7 of the 10 stars. Although it has been shown to be an actual cluster, it appears to many observers as nothing more than an asterism.

NGC 5272	Messier 3	13^h 42.2^m	+28° 23′	GC
6.2 m	⊕ 18.6′		VI	Easy

This really is a splendid globular cluster, easily seen in binoculars and a good test for the naked eye. If using giant binoculars with perfect seeing, some stars may be resolved. A beautiful and stunning cluster in telescopes, it easily rivals Messier 13 in Hercules. It definitely shows pale colored tints, and reported colors include, yellow, blue, and even green; in fact, it is often quoted as the most colorful globular in the northern sky. It is full of structure and detail, including several dark and mysterious tiny dark patches. Many of the stars in the cluster are also variable. This is one of the three brightest clusters in the northern hemisphere and deserves to be on every observers list. The cluster is the prototype for Oosterhoff Type I objects, which are considered "metal-rich." That is, for a typical (is there such an object?) globular cluster, Messier 3 has a relatively high abundance of heavier elements. Located at a distance of about 34,000 light years it is believed to be about 8 billion years old.

Notes

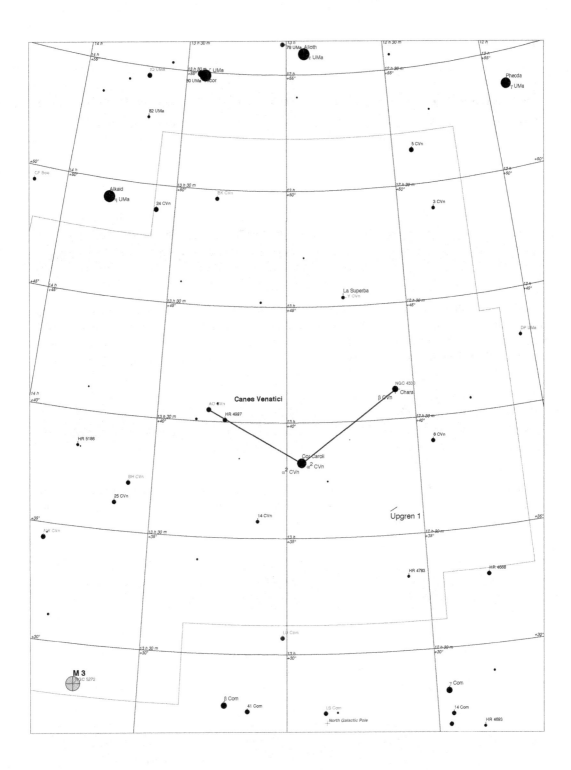

Canis Major

Fast Facts

Abbreviation: CMa	Genitive: Canis Majoris	Translation: The Greater Dog
Visible between latitudes 60° and −90°		Culmination: Dec/Jan

Clusters

NGC 2204	Herschel VII-13	06ʰ 15.5ᵐ	−18° 40′	OC
8.6 m	⊕ 13′	80	III 3 m	Difficult

A slightly difficult cluster to locate and observe, composed of many faint stars but with a nice orange star at its northern limit. This is one of the furthest clusters from the galactic plane, at about 3,000–4,000 light years below it. When you have had enough of food and relatives and television on Christmas Day, then go out and observe this, as it will be at its highest in the sky on that night.

NGC 2243	Collinder 98	06ʰ 29.6ᵐ	−31° 16′	OC
9.4 m	⊕ 5′	100	I 2 r	Moderate

With a 20 cm telescope, the cluster will appear as a faint glow and increasing aperture only shows a few more stars set among a hazy background. Indeed, several observers report that it looks more like a globular cluster than an open cluster! What do you think?

Collinder 121	–	06ʰ 29.7ᵐ	−31° 17′	OC
2.6 m	⊕ 50′	20	IV 3 p	Easy

Here we have a very large open cluster, but one that is difficult to locate due to the plethora of stars in the background. At the northern border of the cluster is o Canis Majoris, best seen with large binoculars or low-power telescopes. Also classified as the Collinder 121 OB Association.

NGC 2287	Collinder 118	06ʰ 46.0ᵐ	−20° 45′	OC
4.5 m	⊕ 38′	70	II 3 m	Easy

Also known as Messier 41, this open cluster is easily visible to the naked eye on very clear nights as a cloudy spot slightly larger than in size than the full Moon. Nicely resolved in binoculars, it becomes very impressive with medium aperture, with many double and multiple star combinations. It contains blue B-type giant stars as well as several K-type giants. Current research indicates that the cluster is about 200 million years old and occupies a volume of space about 25 light years in diameter.

NGC 2354	–	07ʰ 14.3ᵐ	–25° 42′	OC
6.5 m	⊕ 20′	60	III 2 r	Moderate

One really needs moderate and larger apertures to observe this open cluster. In a telescope of about 25 cm, the cluster will appear irregular set among the background field stars of the Milky Way. Increasing aperture will show many features such as star chains, voids, and many more cluster members.

Collinder 132	–	07ʰ 14.4ᵐ	–31° 10′	OC
3.6 m	⊕ 95′	25	III 3 p	Easy

This is one of those clusters that are best seen with binoculars or low magnification. It is a large group of stars and will appear as a richer part of the southern Milky Way. Well worth observing.

NGC 2360	Collinder 134	07ʰ 17.7ᵐ	–15° 38′	OC
7.2 m	⊕ 13′	80	II 2 m	Easy

A beautiful open cluster, Caldwell 58 is irregularly shaped and very rich, sometimes known as Caroline's Cluster. There are many faint stars, however, so the cluster needs moderate-aperture telescopes for these to be resolved, although it will appear as a faint blur in binoculars. This is believed to be an old cluster with an estimated age of around 1.3 billion years, and is intermediate in appearance between the rich open clusters and sparse globulars (or "semi-globular," as such objects are often named). This is what astronomer James Mullaney has to say about the cluster: "Of the many beautiful stellar jewel boxes scattered about the sky, my favorite one, NGC 2360, is neither bright nor glittering or splashing as so many of them are. But to me as a lifelong stargazer, it has always had a lure about it that's hard to explain to someone until they actually see it for themselves. I named it 'Caroline's Cluster' years ago to honor Sir William Herschel's devoted sister who discovered it." This cluster is a little known highlight of the winter sky and well worth seeking out.

NGC 2362	Collinder 136	07ʰ 18.7ᵐ	–24° 57′	OC
4.1 m	⊕ 8′	60	I 3 p n	Easy

Now for a very nice open cluster, known as Caldwell 64, that is tightly packed and easily seen. With small binoculars the glare from Tau (τ) Canis Majoris tends to overwhelm the majority of stars, although it itself is a nice star, with two bluish companion stars (recent research indicates that the star is a quadruple system). But the cluster becomes truly impressive with telescopic apertures; the bigger the aperture, the more stunning the vista. It is believed to be very young – only a couple of million years old – and thus has the distinction of being the youngest cluster in our galaxy. Contains many O- and B-type giant stars, and recent research suggests that as much as 12 % of the stars in this cluster may still have circumstellar discs around them, formed when the stars collapsed from the interstellar gas and dust.

NGC 2367	Collinder 137	07^h 20.1^m	−21° 53′	OC
7.9 m	⊕ 3.5′	30	IV 3 p	Moderate

This open cluster can be glimpsed as a faint indistinct glow with small apertures of, say, 10 cm, but using a larger aperture will show a nice amount of detail, and with many more stars becoming visible. A pleasant cluster to seek out and observe.

Haffner 6	–	07^h 20.1^m	−13° 08′	OC
9.2 m	⊕ 4′	60	IV 3 p	Difficult

Now for the first of two observing challenges (along with Haffner 8, see later). Even in a large aperture telescope, of say, 40 cm, this will only just appear as a faint misty glow of barely resolved stars set against an even fainter haze.

Haffner 8	–	07^h 23.4^m	−12° 20′	OC
9.1 m	⊕ 4.2′	30	IV 3 m	Difficult

Now for the second of our observing challenges, along with Haffner 6, see above. Once again, apertures of at least 40 cm will be needed here in order to see any detail in this small, irregularly shaped and loosely bound open cluster. As with all faint deep-sky objects, a dark sky and perfect seeing will help observations. However, even if your telescope has a smaller aperture than the one quoted here, then by all means give these two clusters a go. Who knows? You may be lucky and see these elusive objects.

Collinder 140	–	07^h 23.9^m	−32° 12′	OC
3.5 m	⊕ 42′	30	III 3 p	Easy

Visible to the naked eye as a richer section of the Milky Way, and likened to the "tuft" at the end of the dog's tail, this is a very large open cluster. Research suggests that it is about 20 million years young, and may even have been created from the same interstellar cloud that formed NGC 2516 and NGC 2547. Located in the cluster is a nice double star Dunlap 47, a widely separated (99″) yellow and blue double of 7.6 and 5.5 magnitudes, respectively, at a position angle of 342°. This region of the sky is wonderful to just leisurely scan with binoculars. Take your time here, as it is worth it.

NGC 2374	Collinder 139	07^h 24.0^m	−13° 16′	OC
8.0 m	⊕ 19′	35	II 3 p	Moderate

An often overlooked open cluster set against the star-rich background of the Milky Way, it is surprisingly easy to see in telescopes of 20 cm aperture and larger. It is quite large, with an irregular shape. It can be seen in large binoculars if averted vision is used.

NGC 2383	–	07^h 24.7^m	−20° 57′	OC
8.4 m	⊕ 6′	40	II 3 m	Difficult

Using a small aperture, this open cluster can be seen as a faint indistinct glow. But it will be small and, admittedly, faint. However, its main claim to fame is that it can be seen in the same field as NGC 2383 (see below) when using a low power.

Ruprecht 18	–	07^h 24.8^m	−26° 13′	OC
9.4 m	⊕ 4′	40	III 1 m	Moderate

The Ruprecht objects are not often mentioned in observing guides, which is a pity, as many of them are worth seeking out. Ruprecht 18 is one such object. In apertures of about 20 cm it will appear as a faint but compact object set among a rich star field. It may even appear slightly nebulous. Increasing magnification will inevitably show fainter stars along with the usual unresolved background glow.

NGC 2384	–	07^h 25.2^m	−21° 01′	OC
7.4 m	⊕ 2.5′	15	IV 3 p	Difficult

This is the more easily seen open cluster than that of its close companion, NGC 2383. In small telescopes it will appear faint and hazy, but with a few resolved stars. Increasing aperture will show more cluster members, but it can appear as if it is part of NGC 2383, albeit separated. Both NGC 2383 and NGC 2384 are difficult to observe, not so much due to their magnitudes but rather to their small size.

Ruprecht 20	–	07^h 26.7^m	–28° 49′	OC
9.5 m	⊕ 10′	30	III 2 m	Difficult

This is a faint object and so will need moderate aperture telescope of 30 cm and more in order for detail to be resolved. In such a telescope, the open cluster will appear as a scattering of faint stars set among a background glow of barely resolved field stars.

Notes

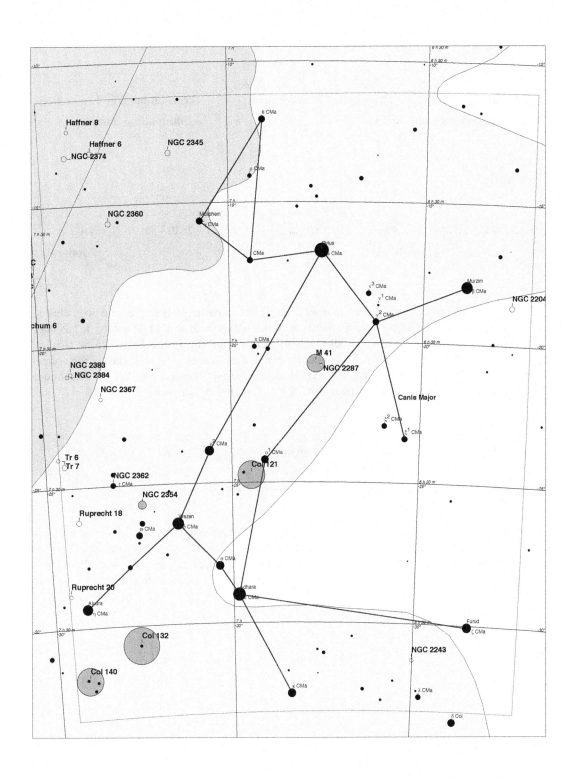

Canis Minor

Fast Facts

Abbreviation: CMi	**Genitive: Canis Minoris**	**Translation: The Lesser Dog**
Visible between latitudes 85° and −75°		**Culmination: January**

Clusters

NGC 2394	–	07^h 28.7^m	+7° 05′	OC
−m	⊕ 10′	40	–	Moderate

Even though this object can be seen in binoculars, it isn't a particularly impressive open cluster. In fact, there seems to be some sort of confusion in the literature as to its real nature; it has been suggested the cluster isn't a cluster at all but is in fact an asterism. In the *Revised New General Catalogue* (Sulentic and Tifft 1973), it is classified as a nonexistent object. Whatever the correct description, observers report it as being "S"-shaped and consisting of several 10th and 11th magnitude stars surrounded by a few fainter members. Check it out and decide for yourself.

Dolidze 26	–	07^h 30.1^m	+11° 54′	OC
−m	⊕ 23′	60?	I V 1 p	Difficult

This open cluster will need a large aperture to be really appreciated. It consists of a large irregular cluster in a roughly oblong shape. The star 6 Canis Minoris may, or may not, be a member. There are several 10th and 11th magnitude stars, more 12th magnitude, and even more faint stars. Both these clusters in Canis Minor are worth seeking out, as they are the only two that can be seen in this area with amateur instruments.

Notes

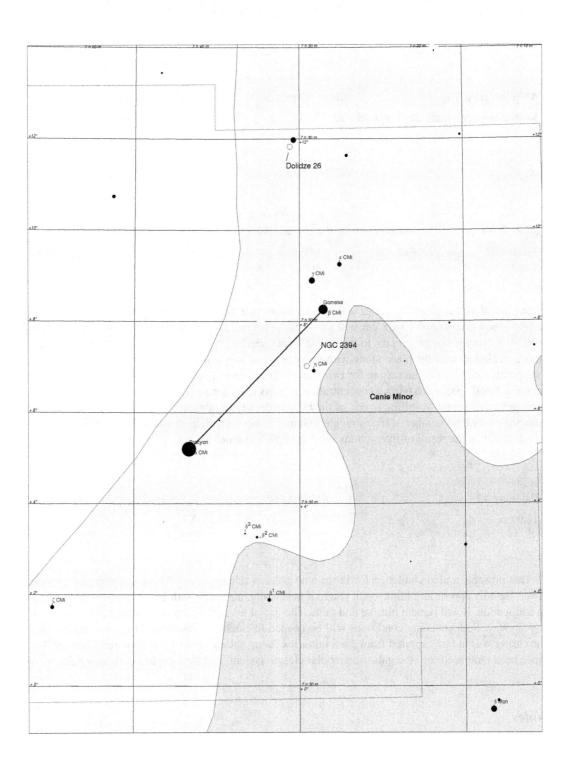

Capricornus

Fast Facts

Abbreviation: Cap	**Genitive: Capricorni**	**Translation: The Sea Goat**
Visible between latitudes 60° and −90°		**Culmination: August**

Clusters

NGC 7099	Messier 30	21^h 40.4^m	−23° 11′	GC
7.4 m	⊕ 12′		V	Easy

In binoculars this globular cluster will appear simply as a tiny, round, hazy patch of light, and even in telescopes of aperture 20–25 cm will show just a bright, asymmetrical core with an unresolved halo. However, as one progresses to larger and larger apertures, this becomes a splendid object with stars speckled all over the place along with many arcs and loops of star chains. From an astrophysics viewpoint, the cluster is interesting for two reasons. It contains a "collapsed" core, so that its central region is tightly packed, having a concentration of mass at its core of about a million times the Sun's mass per cubic parsec, resulting in one of the highest density regions in the entire Milky Way. It also contains a very high number of blue stragglers (stars that are bluer than should be in a globular cluster), which could be the result of interactions due to the high star density.

Palomar 12	–	21^h 46.6^m	−21° 15′	GC
11.99 m	⊕ 3′		XII	Difficult

This presents a nice challenge for those who possess telescopes of 20 cm aperture and greater. It will initially appear as a faint, small patch of nebulosity, and even with an increase of aperture and magnification, it will remain elusive and faint. The use of averted vision adds little, but is worth trying anyway. Perfect seeing conditions will be needed to catch this one. Recent studies suggest that the cluster was in fact captured from the Sagittarius dwarf galaxy about 1.7 billion years ago and has since been embraced into the galactic globular cluster system, and thus the Milky Way's halo.

Notes

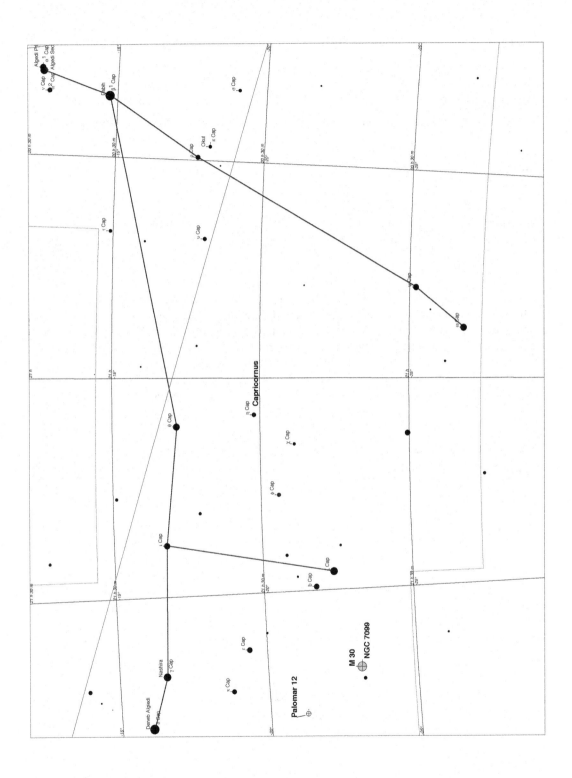

Carina

Fast Facts

Abbreviation: Car	**Genitive: Carinae**	**Translation: The Keel**
Visible between latitudes 20° and –90°		**Culmination: Jan./Feb.**

Clusters

NGC 2516	Collinder 172	07^h 58.0^m	–60° 45′	OC
3.8 m	⊕ 22′	100	I 3 r	Easy

Sometimes called the "Diamond Cluster," this is a lovely object with over 100 stars. Also known as Caldwell 96, it is quite large, spanning nearly half a degree. What's really nice is that there are many double and colored stars sprinkled throughout, including one, near the center of the cluster, that has a distinct red hue to it. The range of magnitudes is impressive, too, ranging from 5th to 11th and even fainter magnitudes. This is one of those clusters that can be observed in all apertures, even with the naked eye. In a small aperture of say 10 cm, the brighter members are resolved, with an unresolved background. Increased aperture brings more stars into sight. This is a true gem of the southern skies.

NGC 2808	–	09^h 23.05^m	–64° 52′	GC
6.2 m	⊕ 14′		I	Moderate

An impressive cluster in moderate to large apertures, it can nevertheless be glimpsed in binoculars, and even a telescope viewfinder, as a small, indistinct blob. With higher magnification and aperture, a nice compact center can be resolved, surrounded by a fainter haze of outlying members. There is considerable debate about the cluster. Did it, like so many other globular clusters, form at the same time as the Milky Way, creating just one generation of stars? Research suggest otherwise, as it seems to contain three generations of stars. Another idea is that it may actually be the remnant of a dwarf galaxy that collided with the Milky Way. Finally, several observers state that its quoted magnitude is incorrect, and should be about 7.8 magnitudes. What do you think?

NGC 3114	–	10^h 02.7^m	–60° 06′	OC
4.2 m	⊕ 35′	171	II 3 r	Moderate

An open cluster that can be seen with the naked eye, NGC 3114 is a large and rich object. Even in a small telescope of, say, 20 cm, innumerable stars, well over 100, can be seen looping in chains and arcs, with several nicely colored members including red and yellow stars. With a larger aperture the cluster will fill the field. However, a question arises: Where does the cluster end and the background disc stars of the Milky Way begin?

NGC 3293	–	10ʰ 35.8ᵐ	–58° 14′	OC
4.7 m	⊕ 5′	90	I 3 r	Moderate

Thought by many to be one of the finest celestial objects in the southern sky, the "Gem Cluster" is a splendid object in moderate to large apertures. In a 20 cm telescope, around 45 stars, several with distinct colors can be seen, including an impressive lovely string of three stars, yellowish, greenish-blue and finally deep orange. Increasing aperture reveals many more stars, including a couple of surprises. The cluster is actually within a very faint nebula, only glimpsed under perfect conditions, along with an even smaller dark nebula lying in the southwest of the cluster. All in all, this is a wonderful object.

IC 2602	–	10ʰ 42.9ᵐ	–64° 24′	OC
1.9 m	⊕ 100′	60	I 3 r	Easy

Best seen in binoculars or a rich-field telescope, the "Southern Pleiades" is a delightful object. It can be glimpsed with the naked eye, as a seemingly separate glow away from the Milky Way, but just south of the Eta Carinae nebula. Increasing aperture will fill the field with a star strewn vista of blue-white and yellow stars. The cluster will appear as two distinct groups, with the 2.8 magnitude star, Theta (θ) Carinae, just east of center. Research suggests it is a very young cluster, of around 30 million years, just half the age of its northern cousins. While you are looking at IC 2602, spend a few minutes looking at the cluster Melotte 101, just 45 arcmin south of Theta (θ) Carina. This is an 8th magnitude haze of about 40 faint stars.

NGC 3532	Collinder 238	11ʰ 05.5ᵐ	–58° 44′	OC
3.0 m	⊕ 50′	150	II 3 r	Easy

Also known as Caldwell 91, Firefly Party Cluster, Pincushion Cluster, or Wishing Well Cluster (take your choice!), this is a superb object (common name notwithstanding). It can be seen with the naked eye seemingly as a part of the Milky Way, but in binoculars it is spectacular, with over 60 stunning stars visible among a background haze. Increasing aperture will show multicolored curved chains and rows of delicate blue, greenish, yellow, orange and red stars set among the dark lanes and unresolved haze of the Milky Way. Even larger apertures, say, of 30 cm and greater, will show stunning views with a seemingly endless panorama of stars. Amazing!

NGC 3603	–	11^h 15.1^m	−61° 16′	OC
9.1 m	⊕ 3.0′	44	III 3 m	Moderate

After the magnificence of the above cluster, a return to the mundane, but nevertheless very interesting open cluster NGC 3603. It lies within the much larger nebulosity known as RCW 57. In moderate aperture, both nebula and cluster will appear as small hazy patch, the stars only becoming resolved with an increase in aperture. However, what makes the cluster an object of intense study is that it is believed to be a starburst region, an area of space with an abnormally high rate of star formation. Located within the cluster are three prominent Wolf-Rayet stars with the largest of the three, NGC 3603-A1, being a blue double star. Their combined masses are believed to be 200 times more dense than our Sun, making (A1-a) one of the largest known stars in the Milky Way with an estimated mass of 116 solar masses, while its companion (A1-b) has a mass of 89 solar masses.[2]

IC 2714	–	11^h 17.3^m	−62° 43′	OC
8.2 m	⊕ 15′	100	II 2 r	Moderate

An often-overlooked cluster, this will show in binoculars as a faintish, small and circular glow. Increasing aperture and magnification presents a nice group of 11th and 12th magnitude stars in arcs and chains with a definite gap between the stars. Also within the field of view is the much fainter and smaller open cluster Melotte 105.

Notes

---

[2]At the time of writing. No doubt this will change as new, more massive stars are found.

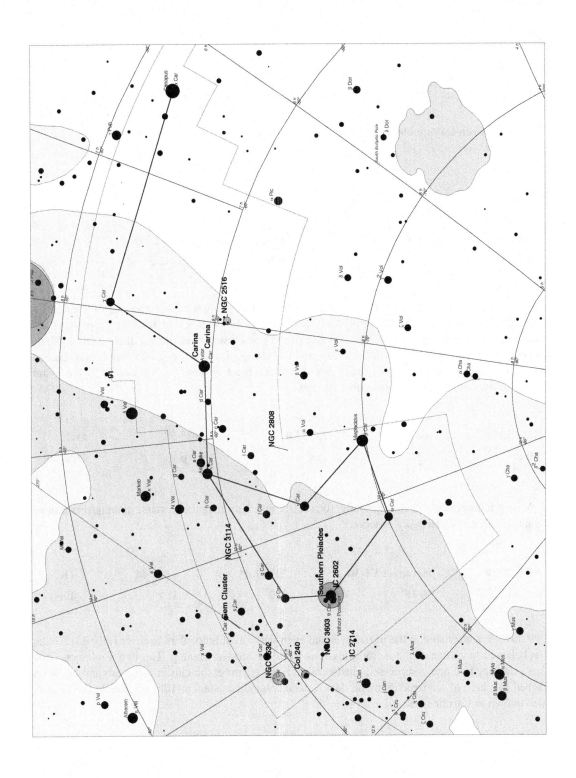

Cassiopeia

Fast Facts

Abbreviation: Cas	Genitive: Cassiopeiae	Translation: Queen of Ethiopia/Andromeda's Mother
Visible between latitudes 90° and −20°		Culmination: October

Clusters

NGC 7654	Collinder 455	23ʰ 24.2ᵐ	+61° 35′	OC
6.9 m	⊕ 12′	100	I 2 r	Easy

Also known as Messier 52, this is a small, rich, and fairly bright cluster and one of the densest north of the celestial equator. Several stars are visible in binoculars, but telescopic apertures are needed to fully appreciate this cluster. It is one of the few clusters that show a distinct color. Many observers report a faint blue tint to the group, and this along with a fine topaz-colored (blue) star and several nice yellow and blue stars make it a very nice object to observe. Apparently, it has a star density of the order of 50 stars per cubic parsec!

King 12	–	23ʰ 53.0ᵐ	+61° 58′	OC
9.0 m	⊕ 2′	15	I 2 p	Difficult

A very faint cluster containing many 10th, 11th, and 12th magnitude stars, making this a most definite challenge to deep-sky observers.

NGC 7789	Herschel VI-30	23ʰ 57.0ᵐ	+56° 44′	OC
6.9 m	⊕ 15′	300	II 1 r	Easy

Visible as a hazy spot to the naked eye, and even with small binoculars is never fully resolvable; it is believed to be one of the major omissions from the Messier catalog. Through a telescope it is seen as a very rich and compressed cluster. With large aperture, the cluster is superb and has been likened to a field of scattered diamond dust. Contains hundreds stars of 10th magnitude and fainter. Also known as Caroline's Rose.

NGC 7790	Herschel VII-56	23ʰ 58.4ᵐ	+61° 13′	OC
8.5 m	⊕ 17′	40	III 2 p	Easy

In small telescopes this cluster will appear as a faint hazy blob, maybe with an elongated aspect. Increasing aperture will show many more stars, but they will still remain faint. Under good conditions star chains may become apparent.

NGC 103	–	00ʰ 25.3ᵐ	+61° 21′	OC
9.8 m	⊕ 5′	30	I 1 p	Moderate

Another small and faint open cluster that, with increasing aperture and magnification reveals, not surprisingly, even more faint stars. It does however stand out well from the background stars.

NGC 129	Herschel VIII-79	00ʰ 29.9ᵐ	+60° 14′	OC
6.5 m	⊕ 21′	35	IV 2 p	Easy

This is a bright and open cluster that is irregularly scattered and uncompressed, making it difficult to distinguish from the background. Up to a dozen stars can be seen with binoculars, but many more are visible under telescopic aperture. Under good observing conditions and using averted vision, the unresolved background stars of the cluster can be seen as a faint glow. In larger telescopes it can look quite magnificent.

NGC 133	Collinder 3	00ʰ 31.2ᵐ	+63° 22′	OC
9.4 m	⊕ 7′	5	IV 1 p	Moderate

Even though this object has been designated an open cluster, this may be a tad enthusiastic, considering it has only a handful of stars. In a small aperture, say, 10 cm, only four stars are seen, and even with an increase in telescope size, only a handful of fainter stars are seen. This is probably more of an asterism.

NGC 136	Herschel VI-35	00ʰ 31.5ᵐ	+61° 32′	OC
11.3 m	⊕ 1.2′	20	II 2 p	Difficult

This is a very nice object, as it is a very small cluster looking like a tiny sprinkling of diamond dust. Although it can be observed with a 15 cm telescope, it needs a very large aperture of at least 20 cm to be fully resolvable. Also, there are many other clusters in this region, so identification can be difficult.

King 14	–	00ʰ 31.9ᵐ	+63° 10′	OC
8.5 m	⊕ 7′	20	III 2 p	Moderate

Often overlooked, this cluster is a faint but rich object. With a 10 cm aperture telescope, several stars can be resolved set against a faint glow.

NGC 146	Collinder 5	00^h 33.1^m	+63° 18′	OC
9.1 m	⊕ 6′	20	IV 3	Moderate

This is a scattered and faint group of about 20 stars when viewed with a small aperture telescope, consisting of both bright and fainter members. An increase in both aperture and magnification will reveal many more, maybe 50 or 60, stars.

NGC 225	Herschel VIII-78	00^h 43.6^m	+61° 46′	OC
7.0 m	⊕ 15′	15	III 1 p n	Easy

A small, open cluster that can be glimpsed in binoculars, it is quite nice in a small telescope. At higher magnification it tends to meld into the rich Milky Way background. However, under perfect conditions, and depending on your eyesight, you may be able to see that the cluster is split into two distinct regions.

NGC 381	Herschel VIII-64	01^h 08.3^m	+61° 35′	OC
9.3 m	⊕ 6′	50	III 2 p	Moderate

A faint cluster, but one that is rich and compressed. Can be resolved with an aperture of 10 cm, but with medium aperture, of say, 20–25 cm, over 60 stars of 12th and 13th magnitude become visible. Under perfect, and I mean perfect, conditions, it may just be glimpsed in binoculars.

NGC 436	Herschel VII-45	01^h 15.6^m	+58° 49′	OC
8.8 m	⊕ 5′	30	I 3	Easy

This is a very nice cluster for several reasons. Even though it is small and faint it is nevertheless quite rich when seen in small apertures, having three bright stars at its center, as well as being barely glimpsed in binoculars. It is also about 50′ northwest of the colorful double star Phi (φ) Cassiopeiae. Increasing both magnification and aperture will resolve many fainter stars in the cluster, many appearing to follow arcs, as well as appearing to be doubles.

NGC 457	Collinder 12	01^h 19.1^m	+58° 20′	OC
6.4 m	⊕ 13′	80	I 3 r	Easy

This is a wonderful cluster, also known as Caldwell 13, and can be considered one of the finest in Cassiopeia. Easily seen in binoculars as two southward-arcing chains of stars, surrounded by many fainter components. The gorgeous blue and yellow double φ Cass and a lovely red star, HD 7902, lie within the cluster. Located at a distance of about 8,000 light years, this young cluster is located within the Perseus spiral arm of our galaxy. Also known as the Owl Cluster.

NGC 559	Herschel VII-48	01ʰ 29.5ᵐ	+63° 18′	OC
9.5 m	⊕ 4.4′	60	II 2 m	Difficult

A very small, very faint cluster, that even with large telescopes remains barely discernible. A good test for both your optics and your eyes, however.

NGC 581	Collinder 14	01ʰ 33.2ᵐ	+60° 42′	OC
7.4 m	⊕ 6′	25	III 2 p	Medium

Also known as Messier 103, this is a nice rich cluster of stars that is resolvable in small binoculars. Using progressively larger apertures, more and more of the cluster will be revealed (as with most clusters). It has a distinct fan shape, and the star at the top of the fan is Struve 131, a double star with colors reported as pale yellow and blue. Close by is also a rather nice, pale, red-tinted star. The cluster is the last object in Messier's original catalog.

Trumpler 1	Collinder 15	01ʰ 35.7ᵐ	+61° 17′	OC
8.1 m	⊕ 4.5′	20	I 3 p	Difficult

Even with a telescope of 12 cm aperture, this small and tightly compressed cluster will be a challenge.

NGC 654	Herschel VII-46	01ʰ 41.1ᵐ	+61° 53′	OC
6.5 m	⊕ 5′	60	II 3 m	Easy

Easily visible in a finderscope, this cluster appears as a small but rich group, with seven 4th magnitude stars at its southeast edge. It stands out rather well against the background and so is easily defined. Well worth exploring with high magnification and larger aperture.

NGC 637	Herschel VII-49	01ʰ 42.9ᵐ	+64° 00′	OC
8.2 m	⊕ 3.5′	20	I 3 p	Difficult

A faint and very condensed cluster. About 10 stars can be seen with a telescope of at least 10 cm aperture, but many more will remain unresolved.

NGC 659	Herschel VIII-65	01ʰ 44.2ᵐ	+60° 42′	OC
7.9 m	⊕ 5′	40	III 1 p	Easy

This is another faint cluster that will just appear to stand out from the background Milky Way. In fact, it may appear to some observers with telescopes of about 20 cm or more that it is nothing but a richer patch of the Milky Way. Try it for yourself and see whether you see a definite cluster, or nothing more than a brighter, richer area of our galaxy.

NGC 663	Herschel VI-31	01ʰ 46.0ᵐ	+61° 15′	OC
7.1 m	⊕ 16′	80	III 2	Easy

This cluster makes a nice change from all the faint and small ones we have seen, as it is fairly large and bright. It can even be glimpsed in a finder. More astonishingly, there are reports of it being glimpsed with the naked eye! About 30 stars can be seen in a 10 cm telescope, with a definite dark lane cutting across it. However, increasing aperture will reveal more of this oft-ignored object, with many 8th, 9th, and 10th magnitude stars set among many fainter members. It also has many double stars systems included and makes one wonder why Messier didn't include it in his famous list.

Collinder 463	Lund 57	01ʰ 48.3ᵐ	+71° 57′	OC
5.7 m	⊕ 36′	40	III 2 m	Easy

This is a nice cluster that exhibits a pleasing aspect in small to medium telescopes, but really looks splendid in larger instruments. Most of the stars are 8.5 magnitude and fainter, and overall the cluster exhibits a crescent shape.

Stock 2	–	02ʰ 15.0ᵐ	+59° 16′	OC
4.4 m	⊕ 60′	50	III 1 m	Easy

Another undiscovered and passed-over cluster! Wonderful in binoculars and small telescopes it lays 2° north of its more famous cousin, the Double Cluster. At nearly a degree across it contains over 50 8th magnitude and fainter stars. Well worth seeking out.

NGC 1027	Herschel VIII-66	02ʰ 42.7ᵐ	+61° 33′	OC
6.7 m	⊕ 20′	150	III 2 p n	Easy

Although this cluster is bright with many members, it is spread out somewhat, and in fact will sometimes appear as just another part of the greater aspect of the Milky Way. It does however have a nice blue star at its center along with an arc of stars at its north. Larger telescopes will reveal a lot more stars, in arcs and chains. Reportedly it can be glimpsed in binoculars.

Collinder 33	–	02^h 59.3^m	+60° 24′	OC
5.9 p	⊕ 39′	25	II 3 m	Moderate
Collinder 34	–	03^h 00.9^m	+60° 25′	OC
6.8 p	⊕ 26′	25	I 3 p	Moderate

This is an often-ignored pair of clusters, and once seen in a telescope it is easy to see why. Even though they are both large, it will immediately become apparent that to most observers, they are nothing more than concentrated star groupings of the Milky Way, and not in fact individual clusters at all. Are they clusters, or just rich star clouds? Have a look for yourself.

Notes

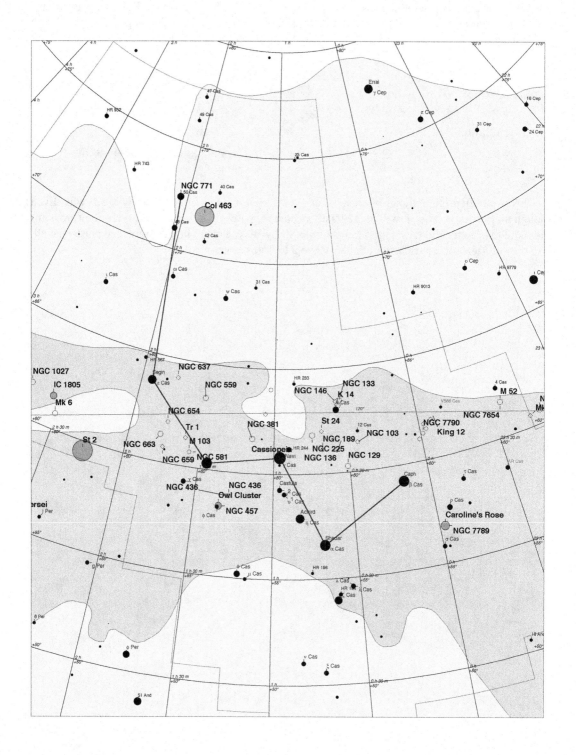

Notes (cont.)

Centaurus

Fast Facts

Abbreviation: Cen	**Genitive: Centauri**	**Translation: The Centaur**
Visible between latitudes 30° and −90°		**Culmination: March**

Clusters

NGC 3680	Collinder 247	11ʰ 25.6ᵐ	−43° 15′	OC
7.6 m	⊕ 12′	50	I 2 m	Moderate

Our first cluster in Centaurus is, like so many others, small and faint, and what makes this one particularly difficult to observe is that it lies in a field devoid of any bright stars. Its brighter members are around 10th magnitude, and even in binoculars will only appear as a small, faint patch of light; however with larger aperture it will begin to show some nicer aspects. So it is worth seeking out.

NGC 3766	Caldwell 97	11ʰ 36.2ᵐ	−61° 37′	OC
5.3 m	⊕ 315	140	I 3 r	Easy

This open cluster should be visible to the naked eye, but can be a problem to locate because it lies in an area of the Milky Way that is already full of hazy patches of stars. Some observers suggest finding the cluster in binoculars first, then having found it, try using your naked eye. However, in small apertures it is a delight, full of colored stars set among a background of fainter white stars. With patience, and a dark sky, many loops and arcs of stars will be seen. But don't try a high magnification on this cluster, as it will cause it to lose some of its appeal.

IC 2944	Collinder 249	11ʰ 36.6ᵐ	−63° 02′	OC
2.9 m	⊕ 65′x40′			Easy

There has been an immense amount of confusion with this object and designation. There are several reasons. Suffice to say that IC 2944 is in fact the Lamda Centauri nebula, whereas Collinder 294 (Caldwell 100) is the open cluster. It is a large grouping of about 40 stars that are embedded in the nebula. In binoculars and small apertures many star streams are visible, but to be honest, the whole cluster is just about eclipsed by the nebulosity. If you are familiar with the glorious image of the nebula within which are the stunning and mysterious Bok globules, then you know what I mean.

Stock 14	–	11ʰ 43.8ᵐ	−62° 32′	OC
6.3 m	⊕ 8′	20	II 3 p	Difficult

Now for another one of the many faint and small open clusters that Centaurus seems to have in abundance. However, there is something of interest here. In a moderate to large aperture, several 8th–20th magnitude stars can be seen, with a dozen or more much fainter stars northeast. However, it is believed that these fainter stars are the true cluster stars, while the brighter stars are just random stars in the field of view.

NGC 5139	Caldwell 80	13ʰ 26.8ᵐ	−47° 29′	GC
3.5 m	⊕ 36′		VIII	Easy

Also known as Omega Centauri, this is a fabulous cluster and one of the showpieces of the night sky. Visible to the naked eye as a clearly seen, hazy patch of light, a stunning sight in binoculars, and a jaw-dropping spectacle in telescopes.[3] This is how the famous American astronomer, Phil Harrington, describes the cluster, "I can still vividly recall my first view of the mighty globular cluster Omega Centauri. Back in 1986, my family and I traveled to Everglades National Park in Florida for a chance to view Halley's Comet. The comet was following a southern trajectory, keeping low in the sky from back home in New York. And while Halley remained disappointing even from that more southerly clime, Omega Centauri proved stunning! The view through my 11x80 binoculars, which clearly resolved many stars throughout, was breathtaking." It contains over several million stars, and some sources put it at having nearly 10 million. It is about 15,800 light years away. Recent research has suggested that the cluster is in fact the central core of a dwarf galaxy that was disrupted by the Milky Way.

NGC 5138	Collinder 270	13ʰ 27.3ᵐ	−59° 02′	OC
7.6 m	⊕ 7′	100	II 2 m	Moderate

Best seen in moderate to large aperture, this is a fairly rich open cluster that stands out quite well from the surrounding background. It consists of around four dozen stars of mostly 11th and 12th magnitude stars, the remainders of which are fainter. Several pairs of stars can also be seen.

Collinder 272	–	13ʰ 30.4ᵐ	−61° 19′	OC
7.7 m	⊕ 10′	100	III 2 m	Difficult

This is a wide, scattered, and poor cluster of 11th magnitude stars. In large binoculars and small telescopes it will appear as mainly a nebulous glow, whereas in a moderate aperture it will appear as an ill-defined group of around ten 10th magnitude stars measuring 5′ by 3′. With a large aperture it can be seen in the same field as the much fainter open cluster Hogg 16, and has been referred to by some observers as a poor man's double cluster.

Trumpler 21	–	13ʰ 32.2ᵐ	−62° 47′	OC
7.7 m	⊕ 5′	20	I 2 p	Difficult

---

[3]Alas, this wonderful globular cluster is not visible from the UK, or northern parts of the United States, but it is truly an amazing object to observe, and should you have the chance to see it, then do so.

Now for a tiny open cluster of about 20 stars, most 12th magnitude and fainter, which can only be appreciated in large aperture equipment. Think of this as a definite test for both your observing skills and telescope optics.

NGC 5286	Caldwell 84	13ʰ 46.4ᵐ	−51° 22′	GC
7.3 m	⊕ 11′		V	Easy

After the disappointment of the preceding clusters, we now have something of a treat. An often-overlooked object, this cluster suffers from being close to the pale orange tinted star M Centauri, but it is a delight using any sort of optical aid. It can just be glimpsed with the naked eye, and in a small telescope it has a slight bluish tinge that contrasts wonderfully with the reddish M Centauri. Increasing aperture will reveal barely resolvable dark areas and arcs of stars.

NGC 5281	Dunlop 273	13ʰ 46.6ᵐ	−62° 55′	OC
5.9 m	⊕ 8′	50	I 3 m	Easy

Another object that is overlooked, this open cluster is a quite bright and splendid sight. It can just be glimpsed with the naked eye, appearing as a 6th magnitude star. Using binoculars it will show as a hazy blob. But where this cluster really begins to shine is in large apertures, because then not only does it consist of a plethora of yellow, bluish, white, and orange stars, but these are set against the now magnified and resolved stunning background of the Milky Way.

NGC 5316	Collinder 279	13ʰ 54.0ᵐ	−61° 52′	OC
6.0 m	⊕ 13′	120	II 2 r	Easy

A fine bright cluster consisting of a few 9th magnitude stars set amongst many more 10th–12th magnitude stars. It is irregular in shape, and with medium to large aperture, many double stars and streams of stars can be glimpsed.

NGC 5460	Collinder 280	14ʰ 07.5ᵐ	−48° 20′	OC
5.6 m	⊕ 35′	60	I 3 m	Easy

Some clusters are small, some are large; this is an example of the latter. This open cluster is enormous, which means it is best observed at low magnification, in order for the entire cluster to be encompassed in a single field. Then there's the fact that it can appear as if it consists of three separate clusters, consisting of about four dozen 9th magnitude stars, and there is the added bonus of the very faint galaxy ESO 221–25 in the background. It is worth seeking out.

NGC 5606	–	14ʰ 27.8ᵐ	−59° 38′	OC
7.7 m	⊕ 3′	15	I 1 p	Difficult

Not only is this a very small cluster, it is also very difficult to find, as it can appear lost in the richness of the background stars. This is one cluster where a medium aperture telescope may struggle, and so inevitably a large aperture telescope will be needed in order to appreciate this object. Once found (if ever!), it will be a small, dense clump of 10th–12th magnitude stars. Naturally, dark and clear skies will be needed.

NGC 5617	–	14ʰ 29.7ᵐ	–60° 42′	OC
6.3 m	⊕ 10′	50	I 3 r	Difficult

Just like its close companion above, NGC 5606, this is another faint and small cluster. However, as it is a tad brighter and larger, it is somewhat easier to locate. It is a large and rich group and under high magnification and aperture, some arcs of stars will be glimpsed. Its claim to fame is that it lies 1.2° west of Alpha (α) Centauri.

Trumpler 22	Collinder 283	14ʰ 31.2ᵐ	–61° 10′	OC
7.9 m	⊕ 7′	50	II 2 p	Difficult

Located close to Trumpler 22 and Alpha Centauri is another well-scattered open cluster. In fact, unless conditions are near perfect, it may be difficult to see this cluster as a separate group from the background stars.

NGC 5662	Collinder 284	14ʰ 35.5ᵐ	–56° 40′	OC
5.5 m	⊕ 30′	50	II 3 r	Moderate

This is a strange cluster in that in rarely appears on observing lists. The reason may be that it is very large, in fact, about the same size as a full Moon, and can be glimpsed in binoculars. Its large size means it is best seen in a low aperture telescope, along with a low magnification. Otherwise the cluster will disappear in the eyepiece. Some observers report that it appears as if the cluster is in two halves, with just a few stars at its center. To make things interesting, there is a nice red star located at the center.

Notes

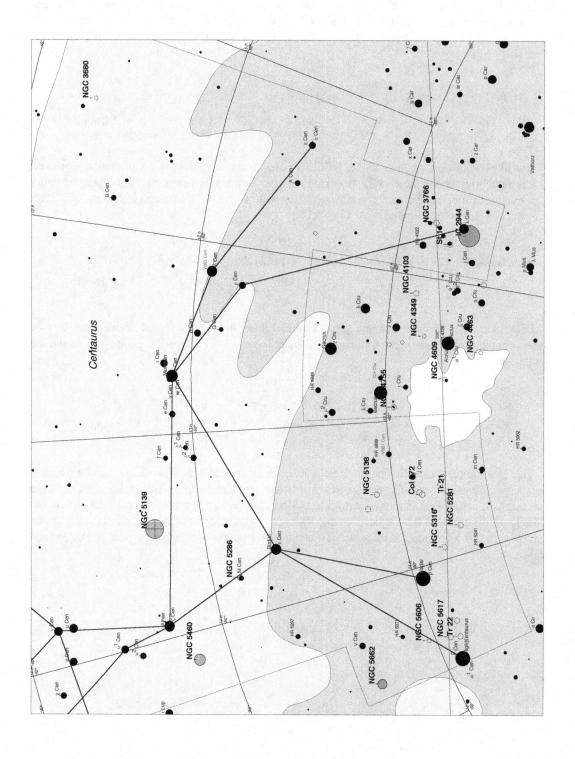

Notes (cont.)

Cepheus

Fast Facts

Abbreviation: Cep	Genitive: Cephei	Translation: King of Ethiopia or Andromeda's Father
Visible between latitudes 90° and −10°		Culmination: September

Clusters

NGC 6939	–	20^h 31.4^m	+60° 38′	OC
7.8 m	⊕ 7.0′	50	I 1 m	Moderate

A moderately bright and small cluster, which is unresolved in binoculars. A challenge, as the brightest member is only of 11.9 magnitude. In telescopes of aperture 10 cm, it will appear as a small hazy spot with just a few very faint stars resolved.

NGC 7023	Collinder 429	21^h 00.5^m	+68° −10′	?
7.1 m	⊕ 5′	–	–	–

You will probably spend an inordinate amount of time searching for this cluster and never discover it. Don't panic, as there isn't one, as the object at this location has erroneously been classified. In fact, there is a nebula here with Collinder 429, to the west of the nebula. It is a very faint, magnitude 13.8, loose cluster, and not worth the effort of looking for unless you have a large aperture telescope. Also known as Caldwell 4.

IC 1396	–	21^h 39.1^m	+57° 30′	OC
3.7 m	⊕ 50′	40	II m n	Easy

Although a telescope of at least 20 cm is needed to really appreciate this cluster, it is nevertheless worth searching out. It lies south of Herschel's Garnet Star and is rich but compressed. What makes this so special, however, is that it is cocooned within a very large and bright nebula.

NGC 7142	Herschel VII-66	21^h 45.9^m	+65° 46′	OC
9.3 m	⊕ 4.3	100	II 2 r	Moderate

In a moderate aperture the cluster will appear as a rich group of around 40 stars ranging in magnitude from 12 to 14, set against a background haze of unresolved stars. Increasing telescope size reveals many more stars belonging to the cluster, making an impressive sight in the eyepiece. What makes the cluster especially interesting however is the nearby reflection nebula NGC 7129. Research indicates that the presence of the nebula indicates that the interstellar cloud could obscure the cluster. A recent study of the Cepheus region concluded that no part of the field might be considered unobscured. Also, determining its age is made problematic due to the obscuration by the nebula. Nevertheless, it is believed to be a very old cluster, which has thrown up its own set of problems, as the cluster possesses a high number of blue stars (blue stragglers), which it should not have if it is old. All in all, a very interesting object.

NGC 7160	Collinder 443	21ʰ 53.7ᵐ	+62° 36′	OC
6.1 m	⊕ 7.0′	50	I 3 p	Easy

Now we have a nice open cluster that can be seen in binoculars. In fact, it could be taken for a globular cluster when seen in large binoculars. In moderate aperture it will appear as a tight group of bright stars. Increasing aperture reveals more of the fainter stars in the group. Well worth a visit.

NGC 7226	Collinder 446	22ʰ 10.4ᵐ	+55° 24′	OC
9.6 m	⊕ 1.8′	25	I 1 0	Difficult

A tiny cluster that is often lost among the background star fields. Large apertures are needed to appreciate this cluster. Averted vision will make a few more members pop out, but it is not impressive.

NGC 7235	Collinder 447	22ʰ 12.4ᵐ	+57° 16′	OC
7.8 m	⊕ 4.0′	30	III 3 p	Moderate

Not a particularly impressive cluster, it consists of a small grouping of around two dozen 9th–12th magnitude stars. With a large aperture more, fainter cluster members can be seen.

NGC 7261	Collinder 450	22ʰ 20.4ᵐ	+58° 05′	OC
8.4 m	⊕ 5.0′	30	III 1 p	Moderate

This isn't a particularly bright or large cluster, perhaps best described as a very loose cluster. Several observers report asterisms shaped like either a trapezium or a "g" as the most prominent feature, with about 15 stars visible in a vague north–south direction. Also, there are reports of a tinge of blue in some of the stars.

NGC 7380	–	22^h 47.3^m	+58° 08′	OC
7.2 m	⊕ 20.0′	50	I 1 m	Easy

Easily seen in binoculars, once found, it will appear as a small, glowing blob of uniform brightness. What becomes apparent however in larger apertures is that there is a definite nebulosity associated with the cluster. Whether dark lanes cut across it, on the limit of visibility, or the cluster is embedded in a barely seen hazy glow, is left for you to decide. Research suggests that the nebulosity is nothing more than the remains of the interstellar material, gas and dust, out of which the cluster formed.

King 19	–	23^h 08.3	+60° 31′	OC
9.2 m	⊕ 6.0′	25	II 2 p	Easy

This is a small open cluster consisting of over a dozen 11th–13th magnitude stars. Its two brightest members have been compared to a pair of eyes!

NGC 7510	–	23^h 11.1^m	+60° 34′	OC
7.9 m	⊕ 7.0′	50	II 2 p	Moderate

A lovely object when using a moderate aperture, consisting of a bright and rich open cluster. The cluster, in any sized aperture, will appear as a wedge-shaped cluster of about 30 stars under a high magnification. The stars in the cluster have a medium range of magnitudes up to around 13th magnitude, and the cluster stands out quite well from the background stars.

NGC 7748	–	23^h 45.0^m	+69° 45′	–

Although many lists have this as a star cluster, it is in fact nothing so exciting but a star. One of the non-existent NGC objects and is included here just to prevent you from searching for it!

NGC 188	Collinder 6	00^h 47.5^m	+14° 5′	OC
7.1 m	⊕ 15′	120	II 2 r	Moderate

One of the oldest open clusters in the sky, if not the oldest, and as such, does not possess any white main-sequence stars, as they have all evolved off the sequence. Not visible in the naked eye, but easily glimpsed with binoculars it will appear as a faint glow about a quarter of a degree across. In any aperture it will appear as a faded scattering of dim stars, and some observers report that it looks like the dying remnants of a cluster, which is quite appropriate, really. High magnification will make the object look less like a cluster and more like a random collection of dim stars. Also known as Caldwell 1.

Notes

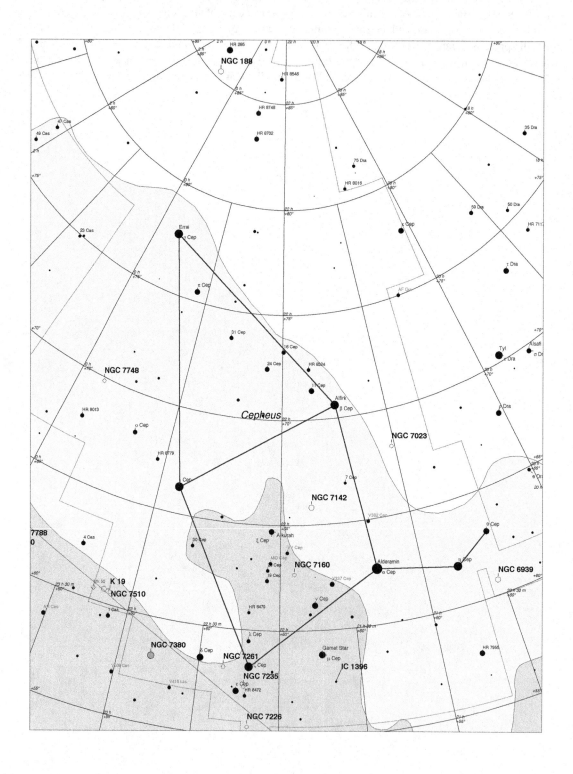

Notes (cont.)

Circinus

Fast Facts

Abbreviation: Cir	Genitive: Circini	Translation: The Compasses
Visible between latitudes 20° and −90°		Culmination: April

Clusters

NGC 5715	Collinder 286	14ʰ 43.5ᵐ	−57° 34′	OC
9.8 m	⊕ 5.0′	30	III 2 m	Moderate

In this very southern constellation, we have a nice open cluster, and in a moderate aperture will consist of around 20 or so 12th–13th magnitude stars. An increase in aperture will naturally resolve many more members of the group, making the cluster look fairly loose and scattered. This isn't a cluster that appears on many observing lists, southerly declination notwithstanding.

NGC 5823	Collinder 290	15ʰ 05.4ᵐ	−55° 37′	OC
7.9 m	⊕ 12′	100	II 2 r	Easy

This cluster can be frustrating to locate, but it's worth persevering over, as once you find it, you are in for a treat! Visible in binoculars, once it's located, Caldwell 88 will look like a piece of the Milky Way, or even the nearby cluster, NGC 5822, in Lupus, that has somehow detached itself. Several stars may hint at resolution, but you are deceived, for these are foreground stars; the cluster's stars are 13th magnitude and fainter. If you are using a small aperture telescope, you may see that NGC 5822 tends to crowd out the fainter NGC 5833, but in large apertures it really is a nice object with nearly 100 stars forming, what some observers imagine, is a teardrop shape. One nice observational item is a small group of about five stars slightly west of the main collection that it stands apart from.

Oddly, in such a rich part of the Milky Way, very little research has been done on the plethora of clusters that can be found in this part of the sky. What work has been done has been confusing to say the least. For instance, it was believed that both clusters were in fact part of a supercluster, or were physically connected in space, but these ideas were proved incorrect. Later, research suggested that NGC 5823 wasn't a cluster at all, but an asterism. But this too was eventually disproved. The latest research in this saga suggests that both objects are separate, distinct open clusters at different distances and with differing ages. The final word is yet to be said on this matter.

Notes

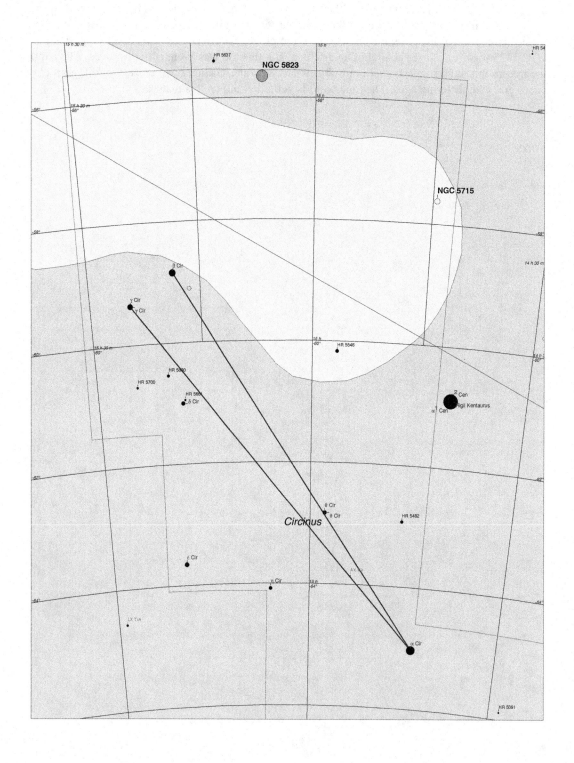

Notes (cont.)

Columba

Fast Facts

Abbreviation: Col	**Genitive: Columbae**	**Translation: The Dove**
Visible between latitudes 45° and −90°		**Culmination: December**

Cluster

NGC 1851	Caldwell 73	05ʰ 14.1ᵐ	−40° 03′	GC
7.1 m	⊕ 12′		II	Moderate

Columba's lone cluster is a globular and is an impressive object. There are reports that it is visible with the naked eye using averted vision, but naturally, only under perfect conditions. It can be glimpsed in binoculars as an out-of-focus star. The cluster lies in a very star poor region of the sky, and consequently, locating the cluster can be a challenge in itself. But it's worth seeking it out. In small apertures of, say, 10 cm, it has been compared in appearance to the head of a comet, with an unresolved core. However, using averted vision, a delicate arm-like structure can appear. Experienced observers mention that the longer you look at it, the more the core becomes partially resolved, and may have a faint yellowish tinge. Larger apertures resolve further detail, a granular-like core surrounded by stars that get fainter as one moves further from the core. This is a very nice object, often overlooked, but once found is a delightful surprise.

Notes

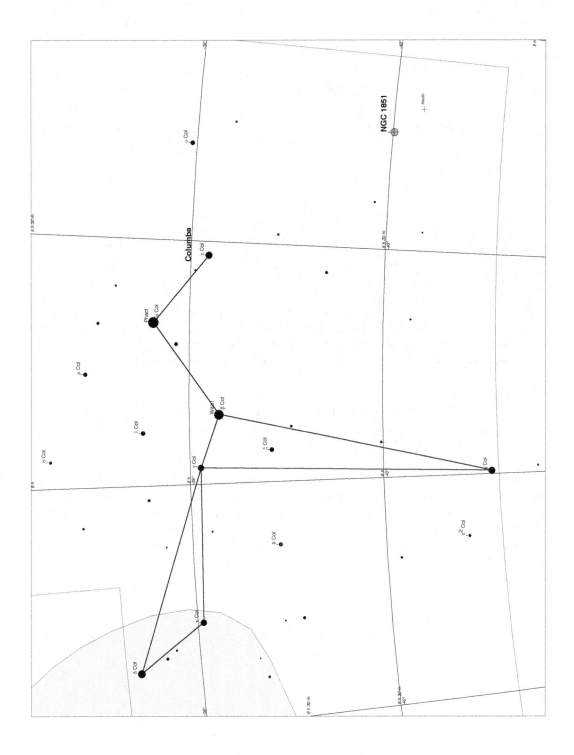

Coma Berenices

Fast Facts

Abbreviation: Com	Genitive: Comae Berenices	Translation: Berenice's Hair
Visible between latitudes 90° and −70°		Culmination: April

Clusters

NGC 4147	Herschel I-19	12ʰ 10.1ᵐ	+18° 33′	GC
10.3 m	⊕ 4.1′		V	Moderate

This is a faint cluster, hazy in appearance with a star-like core. It often presents a challenge for telescopic observers, although it can be seen in apertures as small as 150 mm. Several variable stars are also located within the cluster. Recent research suggests that the cluster was originally a member of the Sagittarius dwarf elliptical galaxy (Sag DEG) but is now being assimilated into the Milky Way!

Melotte 111	Coma Star Cluster	12ʰ 25.6ᵐ	+26° 07′	OC
1.8 m	⊕ 275′	75	II 3 p	Easy

This large and impressive open cluster of 5th and 6th magnitude stars, spanning about 7.5°, is only worth observing with binoculars because telescope observation will lose the clustering effect due to its size on the sky. It can even be glimpsed in a small finder. Believed to be 450 million years old and 260 light years distant, it is the third nearest cluster. Because of its extremely weak gravitational field the cluster may be on the verge of complete disruption. Paradoxically, although this cluster is visible to the naked eye, it has neither a Messier nor an NGC designation.

NGC 5024	Messier 53	13ʰ 12.9ᵐ	+18° 10′	GC
8.33 m	⊕ 13.0′		V	Easy

An often-ignored globular cluster that is a shame, as it is a nice object. It contains about 100,000 stars, none of which are resolved in binoculars, through which it will appear as a faint hazy patch with a brighter center, located in a background star field. Telescopes show a nice symmetrical glow with a concentrated core. Some observers report a colored hue to the cluster. What do you see? It stands up nicely to magnification, and indeed is lovely sight in telescopes of aperture 10 cm and greater. Lies at a distance of around 60,000 light years away from the center of the Milky Way, making it one of the most outlying clusters of the Milky Way.

NGC 5053	Herschel VI-7	13^h 16.4^m	+17° 42′	GC
9.5 m	⊕ 10′		XI	Moderate

A very faint and loose cluster containing only about 3,500 stars, it lies very close to Messier 53, and both can be glimpsed using a low power. However, due to its low surface brightness, dark skies and good seeing are vital, even with large apertures. But, under the right conditions, it is worth seeking out, as it is one of those clusters that are very impressive when seen through a large-aperture telescope. In fact, some observers liken it to an open cluster due to its loose and irregular morphology. Its position in space is also unique, in that it lies about 54,000 light years above the galactic plane. Once again research suggests that the cluster may at one time have been a member of the Sagittarius dwarf elliptical galaxy (Sag DEG), just like NGC 4147 above.

Notes

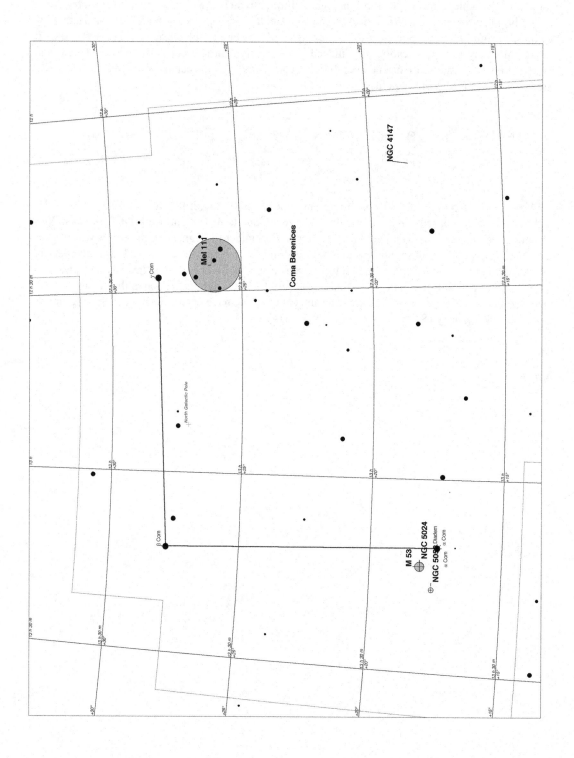

Notes (cont.)

Corona Austrina

Fast Facts

Abbreviation: CrA	**Genitive: Coronae Austrini**	**Translation: The Southern Crown**
Visible between latitudes 40° and –90°		**Culmination: June/July**

Cluster

NGC 6541	**Caldwell 78**	18ʰ 08.0ᵐ	–43° 42′	**GC**
6.3 m	⊕ 13.1′		**III**	**Easy**

Although this cluster suffers slightly from being in a sparsely populated patch of sky, it neverthe-less is a gem. Reported as being just visible to the naked-eye, it can be glimpsed in large finders, and in binoculars appears as just a small and bright hazy glow. Using small apertures, say, 10 cm, the condensed core remains bright, surrounded by a faint halo of dark lanes and star streams. It is quite a breathtaking sight at high magnification. With very large apertures, say 40 cm or more, the core is startlingly bright, with the halo becoming resolved.

Notes

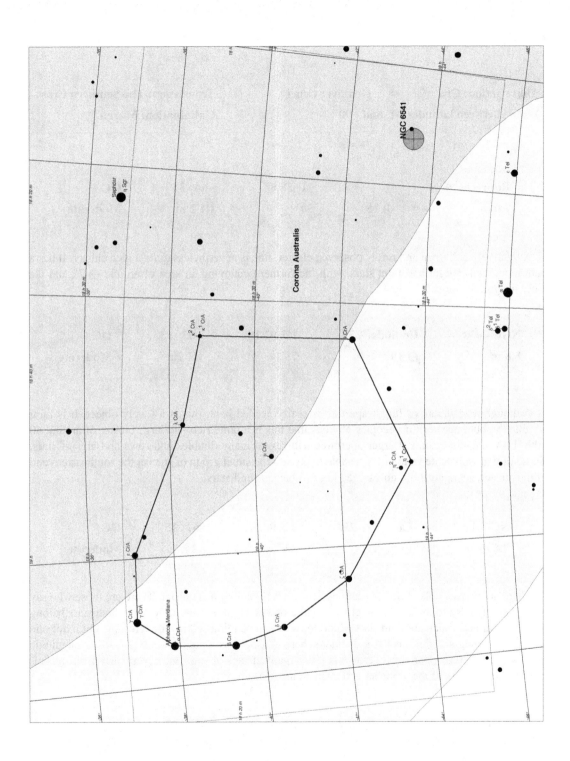

Crux

Fast Facts

Abbreviation: Cru	Genitive: Crucis	Translation: The Southern Cross
Visible between latitudes 20° and −90°		Culmination: March

Ruprecht 98	–	11^h 58.8^m	−64° 34′	OC
7.0 m	⊕ 15′	50	III 2 m	Moderate

A relatively unknown and rarely observed cluster, this is nevertheless quite a nice cluster. It has a both moderately bright and faint stars, with the former occupying an area of around ⊕ 7′, and the latter irregularly spread.

NGC 4052	Collinder 251	12^h 02.1^m	−63° 13′	OC
8.8 m	⊕ 10′	75	III 2 r	Moderate

Although a moderate or larger aperture is really needed here, this is a lovely object. It is faint admittedly, but once seen it becomes obvious that this is a quite rich and large cluster of around 40 10th–13th magnitude stars. Larger apertures will reveal many doubles and rows and arcs of stars. Those of you with acute vision may be able to discern the small group of stars at the southeastern end of the cluster, set against the unresolved haze of background stars.

NGC 4103	Collinder 252	12^h 06.6^m	−61° 15′	OC
7.4 m	⊕ 6.0′	45	I 2 m	Moderate

There are two schools of thought among observers regarding this cluster. There are those who say it is relatively easy to make out the cluster, while others say it is difficult to say which stars belong to the cluster and which are field stars. Reportedly, it can be glimpsed in a finderscope, but it does lie within a stunning star field. It has a distinct spherical aspect with about twenty 9th–14th magnitude stars. What is agreed upon however is that it has many strings of stars and several asterisms, resembling various letters of the alphabet and even an arrowhead!

NGC 4337	Collinder 254	12^h 24.1^m	−58° 07′	OC
8.9 m	⊕ 3.5′	15	II 3 p	Moderate

This is a small, faint and compact cluster that will need at least a moderate aperture to be appreciated. It does have a nice double star about 5′ east of its center. Other than that there isn't much more to say, except it has a lot of 11th–14th magnitude stars. One final note, however – there is some indication that it isn't a true cluster at all but just an apparent group of unrelated stars, i.e., an asterism.

NGC 4349	Collinder 255	12ʰ 24.1ᵐ	−61° 52′	OC
7.4 m	⊕ 15′	200	II 2 m	Easy

After the lackluster appeal of the previous entry, we now have a very interesting object. Visible in binoculars, this is a rich cluster packed with stars, and in larger apertures the field of view is filled with them. Actually, it looks good whatever telescope you use. A member of the cluster is the interesting star NGC 4349–127, a red giant, magnitude 7.4, and orbiting the star is either a Jupiter-like gas giant or brown dwarf nearly 20 times the mass of Jupiter. Its orbit is moderately elliptical, about the same as Mercury's.

NGC 4439	Collinder 259	12ʰ 28.4ᵐ	−60° 06′	OC
8.4 m	⊕ 4.0′	20	II 1 p	Moderate

What is immediately obvious about this cluster is that its brighter stars from a semi-circle, and in the middle of the circle are a string of stars of 9th magnitude. Admittedly the cluster does tend to meld into the background, but once you glimpse it, it is easy to resolve out. Larger apertures and magnifications will show a double star within the semicircle.

NGC 4609	Collinder 263	12ʰ 42.3ᵐ	−62° 59′	OC
6.9 m	⊕ 5.0′	30	II 2 m	Moderate

Also known as Caldwell 98, this is located apparently within the mysterious dark nebulae – the Coalsack. It can be glimpsed with binoculars using averted vision, but it becomes resolved with small apertures. It consists of around six stars aligned north-northeast to south-southeast. Increasing aperture will show about twenty or more 11th magnitude stars. Several observers actually report that at medium magnification and large aperture the cluster resembles a comet. Now, I must make a correction. At the start of this entry we said the cluster is within the nebula; however, this is wrong, as research has shown that the cluster is behind the nebula. Quite a distance behind it actually, at a distance of around 4,300 light years, with the Coalsack at 600 light years. This has the effect of reddening the starlight, or interstellar extinction, as it is called.

NGC 4755	Collinder 264	12ʰ 53.7ᵐ	−60° 21′	OC
4.2 m	⊕ 10′	200	I 3 r	Easy

Considered one the most beautiful and stunning clusters in the entire sky, the Jewel Box, Caldwell 94, or the Kappa Crucis Cluster, as it is also known, is a splendid sight in all apertures. It can be glimpsed with the naked eye, but unlike, say, the Pleiades, it is small and will look like a 4th magnitude star. However, there are a few reports that the two brightest stars in the cluster can be resolved with the naked eye. It's difficult, but apparently it can be done, providing you have perfect conditions and know where to look. It's surprising that such a gorgeous cluster is so small. One would think it would be similar to, say, the Pleiades, but in actuality it is far from it. With binoculars, it will appear as a tiny pyramid of four stars set against an unresolved haze, and any sized telescope will show the true majesty of the cluster. It will display the famous "A," or "Cone"-shaped asterism, but what makes the cluster so special is the amazing range of colors on display. They are yellow, green, orange, white, blue, dark red and reddish gold! It's best not to use a high magnification here, as in doing so the cluster loses its appeal, so use a low magnification to encompass the whole, or most, of the cluster in the field of view. It really is beautiful, with many arcs and chains, along with dark areas. The cluster is one of the youngest known, with an age of only 14 million years.

Notes

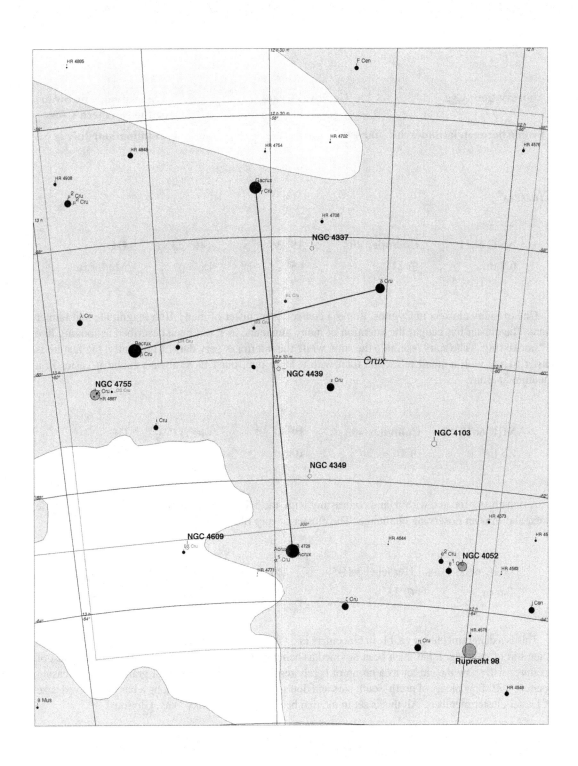

Cygnus

Fast Facts

Abbreviation: Cyg	Genitive: Cygni	Translation: The Swan or the Northern Cross
Visible between latitudes 90° and −40°		Culmination: July

Clusters

NGC 6811	Collinder 402	19ʰ 36.9ᵐ	+46° 23′	OC
6.8 m	⊕ 21′	65	IV 3 p	Moderate

One of many clusters in Cygnus, this is a coarse open cluster of many 10th magnitude and fainter stars. The cluster has caught the attention of many amateurs, as it has been described as looking like a "smoke ring." The stars resemble the ring, while the interior is very dark. Apparently, the feature is easier to see with a small telescope than with a large one, but I have always found it easier in a medium aperture.

NGC 6819	Collinder 403	19ʰ 41.3ᵐ	+40° 11′	OC
7.3 m	⊕ 10′	100	I 1 r	Moderate

A rich cluster located within, and contrasting with, the Milky Way. Contains many 11th magnitude stars, and thus an observing challenge. The cluster is very old at over 3 billion years.

NGC 6866	Herschel VII-59	20ʰ 03.9ᵐ	+44° 09′	OC
7.6 m	⊕ 15′	80	II 2 m	Easy

This is a delightful cluster visible in binoculars as a faint glow that, with averted vision, appears to have some sort of mottling. It has even been reported as being glimpsed in a finderscope. Even the smallest of apertures will show the cluster as a group of barely resolved glow with a hint of granularity. Increasing aperture will show plenty of north–south bars and double and triple stars set among a barely resolved haze of fainter cluster members. All this is set in the rich backdrop of the Milky Way. Glorious!

NGC 6871	Collinder 413	20ʰ 05.9ᵐ	+35° 47′	OC
5.2 m	⊕ 20′	30	II 2 p n	Easy

Here we have a nice cluster that is easily seen in small telescopes. It appears as an enhancement of the background Milky Way. Binoculars will show several stars of 7th–9th magnitude surrounded by fainter members.

NGC 6883	Collinder 415	20^h 11.3^m	+35° 50′	OC
8.0 m	⊕ 35′	30	IV 2 m n	Moderate

This cluster suffers from being in a particularly rich star field, and being faint and large just compounds the difficulty in observing it. What can be seen, however, in moderate to large aperture is that the cluster breaks up into several groups that are crossed by dark lanes.

Collinder 419	Lund 940	20^h 18.1^m	+40° 44′	OC
5.4 m	⊕ 4.5′	60	IV 2 p	Moderate

This cluster lies in the central part of Cygnus, one of the richest parts of the Milky Way. The cluster itself surrounds the double star Struve 2666. Although the integrated magnitude is low, the cluster is poor and spread out, thus it looks faint. Not a particularly impressive object.

NGC 6910	Herschel VIII-56	20^h 23.2^m	+40° 47′	OC
7.4 m	⊕ 10′	60	I 2 m n	Moderate

This is a nice little cluster that has been given the name The Rocking Horse cluster by some observers. It can be seen in a small aperture as a nice grouping with two 7th magnitude stars set among it. With medium and larger aperture, a definite "Y" asterism becomes apparent. An interesting point can be made here: the two stars mentioned earlier, both appear yellow, but the southernmost star, V2118 Cygni is in fact a B1.5Ia supergiant star, which should look blue. So why does it appear yellow? The answer lies in the fact that the cluster lies not far off the galactic plane, and so subject to a significant amount of interstellar extinction and reddening from interstellar dust.

NGC 6913	Collinder 422	20^h 23.9^m	+38° 32′	OC
6.6 m	⊕ 7′	80	I 2 m n	Easy

Also known as Messier 29, this is a very small cluster and one of only two Messier objects in Cygnus. It contains only about a dozen stars visible with small instruments, and even then benefits from a low magnification. However, studies show that it contains many more bright B0-type giant stars, which are obscured by dust. Without this, the cluster would be a very spectacular object.

NGC 7128	Herschel VII-40	20^h 43.9^m	+53° 43′	OC
9.7 m	⊕ 4′	35	I 3 m	Moderate

This open cluster is tiny but compact, which paradoxically makes it easier to see if using averted vision. In medium aperture telescopes at low magnification, only four or five stars will be resolved. Increasing both aperture and magnification will reveal more of the fainter cluster members.

NGC 6996	Collinder 425	20ʰ 56.5ᵐ	+45° 28′	OC/?
10 p	⊕ 5′	20	III 2 p	Difficult

Now we come to a cluster that seems to have caused a lot of confusion in the handbooks. Many seem to confuse this cluster with a nearby cluster NGC 6997. In fact, such is the magnitude of the problem that some star maps, and books, leave either one or both clusters out of the lists. Although there is a nice cluster at the position of NGC 6997, at the position of NGC 6996 there is an admittedly small group of stars, but to call it a cluster is sophistry. Thus, you decide how to describe this object. What do you see?

Collinder 428	–	21ʰ 03.2ᵐ	+44° 35′	OC
8.7 p	⊕ 13′	20	Iv 1 p n	Moderate

Now for a very interesting cluster, or should we say, not a cluster! But first, this object can be found in the northeastern part of the much more famous North American Nebula. It's best to use a large aperture here, and it will appear as a large, very loosely scattered cluster of around 15–20 stars. A prominent group of stars form an asterism in the shape of an "A" at the cluster's center. A distinctly brighter star lies on the western-northwestern edge of the cluster. Now for the interesting part of this story. When this cluster was first identified early in the twentieth century, it was assumed Collinder 428 was an open cluster embedded in the nebula. But research has revealed that the cluster is merely a group of unrelated background stars shining through a hole in the nebula, giving them the appearance of a star cluster.

NGC 7031	Collinder 430	21ʰ 07.2ᵐ	+50° 53′	OC
9.1 m	⊕ 5′	50	III 2 m	Moderate

In a medium aperture telescope the cluster will appear as a small grouping of 10 or so stars in a semi-circular pattern. Depending on local sky conditions, averted vision may, or may not, reveal any fainter stars. Larger apertures, along with large or medium magnification, can reveal a pretty colored pair of orange and blue stars.

IC 1369	Collinder 432	21ʰ 12.1ᵐ	+47° 44′	OC
8.8 m	⊕ 4′	40	II 2 m	Moderate

In a moderate aperture this cluster will appear as a small, circular and rich cluster of very faint stars that can only be resolved with averted vision. When looking directly at the cluster it appears as a very hazy blob of light. Larger apertures will resolve many of its 11th–14th magnitude stars. However, what makes this cluster worth seeking out is the dark nebula Barnard 361 half a degree south. It is quite an impressive sight – a starless gap in the rich background of the Milky Way.

NGC 7039	Collinder 431	21^h 21.8^m	+45° 37′	OC
7.6 m	⊕ 25′	50	IV 2 m	Moderate

Another one of Cygnus's open clusters that is faint and ill defined. In a moderate aperture it will appear as a large, irregularly shaped, loose cluster of faint stars, 11th–12th magnitude that is poorly defined. There is a bright foreground star.

NGC 7062	Collinder 434	21^h 23.4^m	+46° 23′	OC
8.3 m	⊕ 6′	30	II 2 m	Moderate

A faint cluster best seen with moderate to large apertures, although it can be seen with smaller instruments. It benefits from being in a relatively featureless part of the sky, so stands out well. It has a reported elongated shape, but this may be due to one's seeing conditions.

NGC 7063	Collinder 435	21^h 24.3^m	+36° 29′	OC
7.0 m	⊕ 7′	12	III 1 p	Easy

In large binoculars, the cluster will be seen as a very nebula-like patch. Set among a rather rich area of the sky, it doesn't stand out too well, no matter what aperture is used.

NGC 7067	Collinder 436	21^h 24.4^m	+48° 00′	OC
9.7 m	⊕ 3′	–	II 1 p	Difficult

Now for a tiny and faint cluster that hardly deserves the classification as a cluster. It will appear, in moderate and large apertures, as a vague and indistinct group of 11th magnitude and fainter stars.

NGC 7082	Lund 992	21^h 29.2^m	+47° 08′	OC
7.2 m	⊕ 25′	20	IV 2 p	Moderate

Using binoculars the cluster appears as a large but faint hazy blob of starlight. With medium aperture, the cluster will appears as a large, but poorly defined, cluster; spread over a large area of the sky observers have reported that using a larger aperture and magnification, several stars exhibit colors, a few yellow stars and a lone red star.

NGC 7082	Lund 992	21^h 29.3^m	+47° 08′	OC
7.2 m	⊕ 25′	30	IV 2 p	Moderate

This large and ill-defined cluster can be glimpsed in large binoculars as an oval haze. With larger or medium aperture it will appear as a group of five bright stars with some nebulosity intermixed within the stars. Averted vision resolves many faint stars in the background. However, using large aperture will show the cluster to be very nice indeed. At high magnification up to four dozen stars will be seen, and as a bonus there is a nice gold and blue double on its southern south side. Some observers have reported seeing a dark lane branch off from its northern side.

NGC 7086	Herschel VI-32	21^h 30.4^m	+51° 36′	OC
8.4 m	⊕ 9′	50	II 2 m	Moderate

This is a pleasant cluster, quite large, and although not particularly bright, it is dense, making it visible in binoculars, and in fact is nice in all apertures. Small apertures will show a well-scattered cluster of about fifty or more 10th magnitude stars, while larger telescope will show more faint members, and at the northeast of the group is an apparently starless patch of sky.

NGC 7092	Collinder 438	21^h 32.2^m	+48° 26′	OC
4.6 m	⊕ 31′	30	III 2 m	Easy

Also known as Messier 39, this is a nice cluster in binoculars; it lies at a distance of 840 light years. About two dozen stars are visible, ranging from 7th to 9th magnitude. What makes this cluster so distinctive is the lovely color of the stars – steely blue – and the fact that it is nearly perfectly symmetrical, having a triangular shape. There is also a nice double star at the center of the cluster.

IC 5146	Collinder 470	21^h 53.4^m	+47° 16′	OC
–m	⊕ –	20	III 2 p	Moderate

You will find a lot of confusion about this cluster, as the designation IC 5146 is in fact for the nebula – the Cocoon Nebula – and not the cluster. The cluster itself is nondescript, just the group of stars associated with the nebula. The question remains, is the nebula the progenitor of the cluster?

Notes

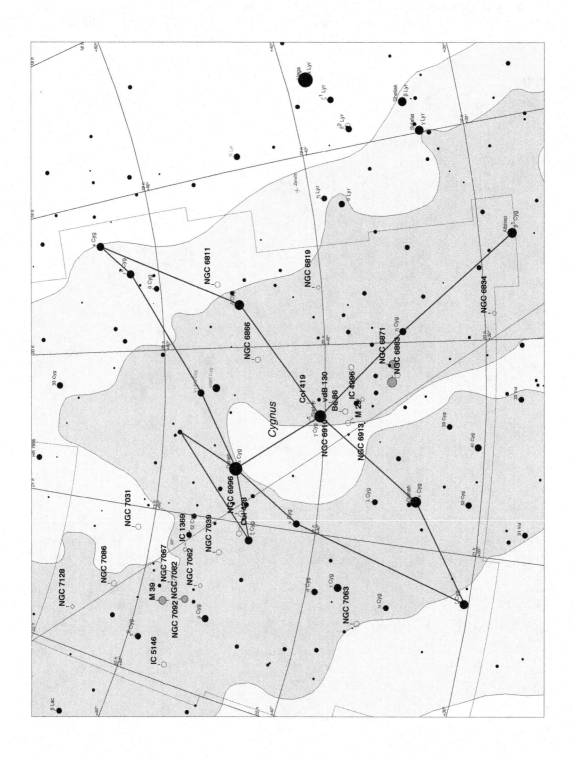

Notes (cont.)

Delphinus

Fast Facts

Abbreviation: Del	Genitive: Delphini	Translation: The Dolphin
Visible between latitudes 90° and −70°		Culmination: August

Clusters

NGC 6934	Caldwell 47	20ʰ 34.2ᵐ	+07° 24′	GC
8.8 m	⊕ 6.2′		VIII	Moderate

This globular cluster is a difficult object even for binoculars, where it will appear as a tiny patch if light. It is just resolvable with a 10 cm aperture telescope as a small, bright and round cluster, with a brighter and condensed center. Some observers report that the use of averted vision aids in seeing some faint structure within the cluster. It has many blue straggler stars, and was one of the first objects to be imaged by the Gemini North Telescope, which resolved its core. In addition, it moves in a retrograde motion to that of the Milky Way's rotation, suggesting that it may be the leftover remnant of a disrupted satellite galaxy.

NGC 6950	–	20ʰ 41.2ᵐ	+16° 38′	OC
−m	⊕ 15′		–	Difficult

This open cluster is a difficult object, for in any aperture and really need at least 30 cm and more to be glimpsed. It is classified as a "non-existent" cluster. When seen at all, it will appear as a large, loose cluster of about 40–50 12th magnitude and fainter stars. Not spectacular by any means but worth seeking out if only as a test of your observational skills. Incidentally, some references have classified this as an asterism, so more work needs to be done, obviously.

NGC 7006	Caldwell 42	21ʰ 01.5ᵐ	+16° 11′	GC
10.6 m	⊕ 2.8′		I	Moderate

Here we have another very distant cluster, at 140,000 light years from the Solar System. In telescopes it appears as a small, unresolved disc, not unlike a planetary nebula. However, even with large binoculars little will be seen unless averted vision is used. A very large and old cluster located far out in the galactic halo.

Notes

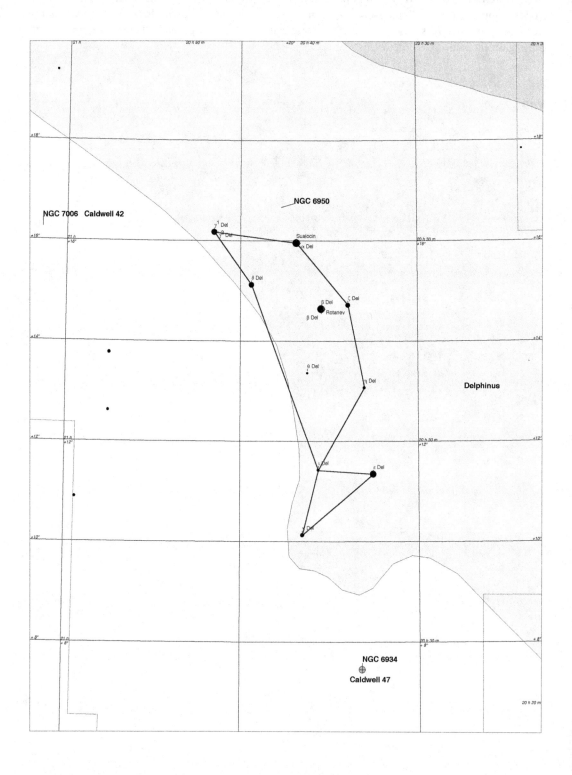

Notes (cont.)

Dorado

Fast Facts

Abbreviation: Dor	**Genitive: Doradûs**	**Translation: The Swordfish**
Visible between latitudes 15° and −90°		**Culmination: December**

Clusters

NGC 1731	–	04^h 53.5^m	−66° 55′	OC/Neb
9.9 m	⊕ 8.0′	–	–	Moderate

Our first cluster in Dorado lies within the Large Magellanic Cloud, LMC, as do most of the clusters in the constellation. With nearly all the clusters that I shall look at, many are visible in binoculars and are often associated with nebulosity, but they are far from us, and so will appear small and faint. Thus to see them in any detail, a medium to large aperture telescope would be better. Nevertheless, always try to observe the cluster in binoculars, and then move on to a telescope. Just remember that when looking at these wonderful objects, you are in fact looking at an object that resides in another galaxy! This open cluster is easily seen and will appear as an irregular, well-scattered group of around a dozen 13th magnitude stars. There is an obvious double star, and, as a bonus, the complete cluster is enveloped in faint, but identifiable, nebulosity.

NGC 1755	–	04^h 55.2^m	−68° 12′	OC
9.9 m	⊕ 2.0′	–	–	Moderate

An easy to locate cluster, as it is the brightest object (in a relative way) in the area, this open cluster will appear as a compact object, relatively bright and circular in shape. In a medium aperture it will not be resolved, and even with a larger telescope only a handful of stars will be seen.

NGC 1761	–	04^h 56.6^m	−66° 28′	OC/Neb
9.9 m	⊕ 1.2′	–	–	Moderate

This is a rich cluster of faint stars of 10th–13th magnitude that melds at its periphery into the background field. But what makes the cluster special is that it is located in a superb and much larger complex of stars and nebulously that is so large and bright it can be glimpsed in a finderscope and is well worth scanning with binoculars.

NGC 1820	–	05ʰ 04.1ᵐ	–67° 16′	OC
9.0 m	⊕ 10.0′	–	–	Difficult

Another large but faint open cluster of over a dozen 10th–13th magnitude stars. Those observers who have large telescopes may be able to see that on its western side are a couple of much smaller and fainter groups of stars. These are in fact the clusters NGC 1814 and NGC 1816. So in fact the whole complex is NGC 1820. There is some very faint nebulosity associated with the smaller NGC 1816.

NGC 1818	–	05ʰ 04.2ᵐ	–66° 26′	GC
9.7 m	⊕ 3.4′		–	Moderate

Another cluster whose pedigree is in question. It has been classified as either an open cluster or a globular cluster, depending on what you read. Current research favors the latter. Visually it is impressive with a bright center enclosed by a fainter region. It is a very young object, maybe 40 million years, compared to the billions of years of similar clusters in the Milky Way.

NGC 1850	–	05ʰ 09.8ᵐ	–68° 55′	OC
9.0 m	⊕ 3.4′	–	–	Easy

You are in for a treat when you observe this object, for not only is it a spectacular object, being the brightest star cluster in the LMC, but it is in a truly fabulous part of the sky that abounds in clusters and nebulae. Don't be fooled into thinking it is a globular cluster, because it isn't! This is a rich, very bright, large and round globular-looking cluster. It also has a very prominent bright center that can't be missed, along with several fainter stars in its partially resolved halo. Research suggests that it is only 50 million years old, with no counterpart in the Milky Way. Within 30′ of the cluster are NGC 1854, NGC 1858, NGC 1856, NGC 1836, NGC 1839, NGC 1847, NGC 1860, NGC 1863 and NGC 1865. All in all, this is a wonderful part of the sky.

NGC 1858	–	05ʰ 09.9ᵐ	–68° 54′	OC/Neb
9.9 m	⊕ 5.0′	–	–	Moderate

An open cluster that is overshadowed by the nebulosity in which it resides, this is a large cluster with an oval shape and a few 12th magnitude stars set against a haze of unresolved stars. To one side is a bright nebulous knot that cannot be missed.

NGC 1866	–	05^h 13.7^m	–65° 28′	GC
9.7 m	⊕ 5.1′		–	Moderate

Lying at the seemingly edge of the LMC is what is believed to be one of the largest and most massive clusters in that galaxy. Like most clusters in the Dorado, one will need a medium to large aperture to really appreciate this object. It is fairly large and bright, although in all except the largest telescopes, resolution of the cluster may not be possible, but a vague mottling can be glimpsed. Research indicates that it is a relatively young object, at about 100 million years, but depending on what star catalog one looks at, or indeed research paper, the question that arises is what type of cluster are we looking at – globular or open?

NGC 1968/ NGC 1974/NGC 1955		05^h 27.4^m	–67° 27′	OC/Neb
9.0 m	⊕ 1.1′		–	Moderate

Another open cluster that is associated with nebulosity, like so many others. This cluster, one of three (including NGC 1974 and NGC 1955), and the central cluster, is a group of stars of around 10th–13th magnitude. NGC 1974 is the easternmost group of about a dozen or more stars. Both clusters stand out against the background haze.

NGC 1984/NGC 1994		05^h 27.6^m	–69° 08′	OC
10.0 m	⊕ 1.0′	–		Difficult

Both of these close clusters will appear as bright and tight knots of light. The former is a slightly elongated haze with only a handful of stars being resolved using the largest aperture, whereas the latter appears as a smaller, circular, unresolved knot.

NGC 1983		05^h 27.7^m	–68° 59′	OC
8.0 m	⊕ 0.6′		–	Difficult

This is a fairly bright but small cluster that will appear as a hazy glow, along with a few stars at its edge that are not actual members of the cluster but rather stars in the field of view. Indeed even with large aperture and high magnification, only around four stars belonging to the cluster will be resolved.

NGC 2004	–	05^h 30.7^m	–67° 17′	OC
9.6 m	⊕ 2.7′		–	Moderate

This is a lovely open cluster that is very bright with an intense core, from which appears streams of stars emanating in all directions. The core will be unresolved, but its outer reaches will show several streams of 12th magnitude and fainter stars.

NGC 2042	–	05^h 35.9^m	–68° 55′	OC/Neb
9.9 m	⊕ 8.5′	–	–	Moderate

A nice rich cluster irregularly scattered in an oblong-shaped area. It has about two dozen 10th–13th magnitude stars set against an unresolved haze. It lies very close to the Tarantula Nebula, about 15′ to the clusters southeast.

NGC 2050	–	05^h 36.6^m	–69° 23′	OC/Ast
9.3 m	⊕ 1.0′	–	–	Moderate

Visually, this will appear as just a locally brighter spot with maybe a dozen stars set against a hazy background glow. It is however embedded at the edge of a remarkable stream of stars that begins close to the south side of the tendrils of the Tarantula Nebula. But is it a real open cluster or, just like so many others, an asterism?

NGC 2055	–	05^h 36.8^m	–69° 29′	OC/Ast
8.5 m	⊕ 0.6′	–	–	Moderate

What we have here is a dilemma. On the one hand we have a group of five or six stars at the coordinates given that are often classified as the cluster, but it is more likely an asterism. However, the great man himself, John Herschel, said that NGC 2024 was a large object, so he must have been referring to the very large star cloud that is nearby. You observe it and you decide. Incidentally, the star cloud is very impressive. The asterism is not.

NGC 2100	–	05^h 42.2^m	–12° 15′	OC
9.6 m	⊕ 2.3′	–	–	Moderate

Often mistaken for a globular cluster due to its appearance, this open cluster is an impressive object. It is slightly irregular in appearance, and its periphery can be resolved into many 12th and 13th magnitude stars with large aperture and high magnification.

Notes

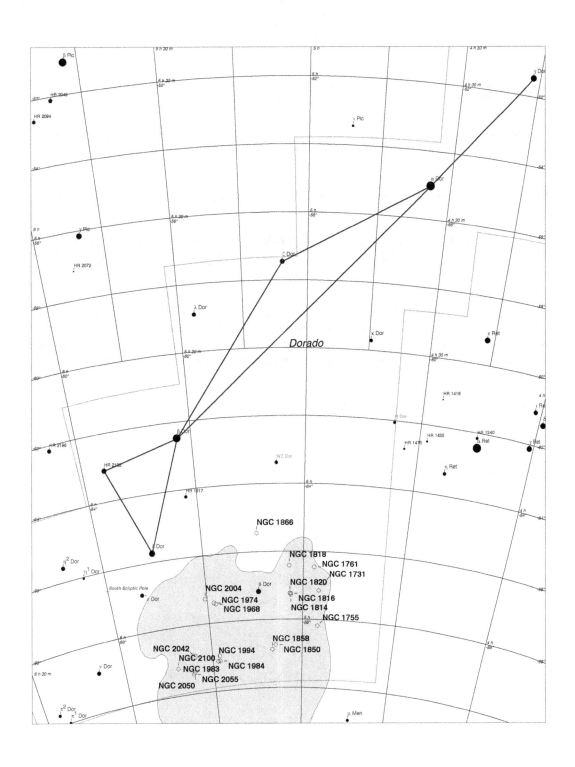

Fornax

Fast Facts

Abbreviation: For	Genitive: Fornacis	Translation: The Laboratory Furnace
Visible between latitudes 50° and –90°		Culmination: November

Cluster

NGC 1049	Hodge 3	02ʰ 39.8ᵐ	–34° 15′	GC
12.6 m	⊕ 0.8′		–	Difficult

This globular cluster is included even though it is at the limit of visibility for the telescope aper-tures usually used, for a very specific reason. This cluster is not a member of the Milky Way, but it is in the nearby Fornax dwarf galaxy, itself a member of the Local Group of galaxies. At a distance of 630,000 light years, it is visible in large aperture, but the parent galaxy will be invisible. It will look faint, small, and circular, and higher magnification and averted vision may improve viewing. There are four other globular clusters in the galaxy, but in order to see these you will need a very large telescope, beyond the scope of this book.

Notes

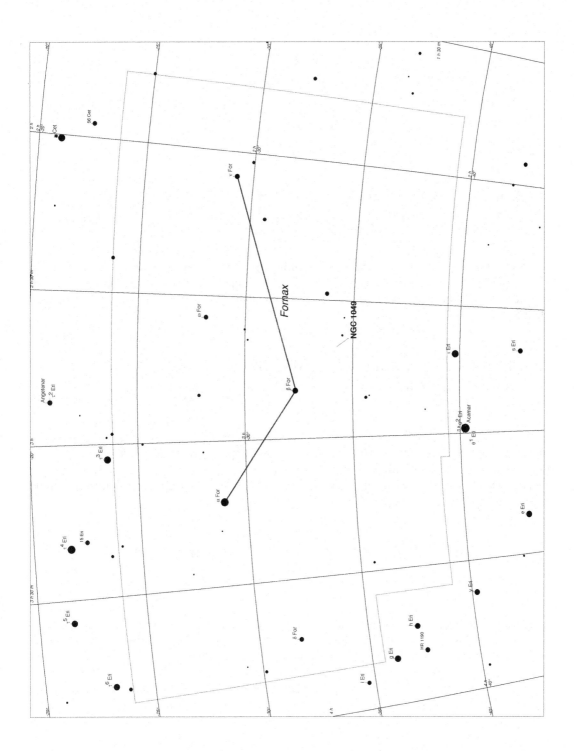

Gemini

Fast Facts

Abbreviation: Gem	Genitive: Geminorum	Translation: The Twins
Visible between latitudes 90° and −60°		Culmination: January

Clusters

NGC 2129	Collinder 77	06ʰ 01.0ᵐ	+23° 10′	OC
6.7 m	⊕ 7′	40	II 3 p	Moderate

This open cluster is visible as a tiny hazy spot in a finderscope, and in a moderate aperture as a small bright and compact grouping of about 10th and fainter magnitude stars surrounding two brighter members. In larger apertures many more faint members are visible, and appear to form small star chains.

IC 2157	Collinder 80	06ʰ 04.0ᵐ	+23° 04′	OC
8.4 m	⊕ 8′	20	III 2 p	Moderate

A faint and small open cluster that is not really impressive in small aperture telescopes, appearing as a random collection of stars. In 10×50 binoculars it will appear as a tiny blur. (Note, however, that in the same field can be seen Messier 35 and NGC 2158.) Using larger aperture will just reveal even fainter stars in the cluster, seemingly in two groupings, with an orangish, 9th magnitude star in the southernmost group.

NGC 2158	Collinder 81	06ʰ 07.5ᵐ	+24° 06′	OC
8.6 m	⊕ 5′	70	II 3 r	Moderate

Lying at a distance of 16,000 light years, this is one of the most distant clusters visible using small telescopes and lies at the edge of the galaxy. It needs a 20 cm telescope to be resolved, and even then only a few stars will be visible against a background glow. It is difficult to say where the cluster ends and the Milky Way begins. It is a very tight, compact grouping of stars, and something of an astronomical puzzle. Some astronomers class it as intermediate object lying between an open cluster and a globular cluster, and it is believed to be about 800 million years old, making it very old as open clusters go.

NGC 2168	Collinder 82	06ʰ 08.9ᵐ	+24° 21′	OC
5.1 m	⊕ 28′	200	III 2 m	Easy

Also known as Messier 25, this is one of the most magnificent clusters in the sky. Visible to the naked eye on clear winter nights with a diameter as big as that of the full Moon, it seems as if the cluster is just beyond being resolved. Many more stars are visible in binoculars set against the hazy glow of unresolved members of the cluster. With telescopes, the magnificence of the cluster becomes apparent, with many curving chains of stars.

Collinder 89	–	06ʰ 18.0ᵐ	+23° 38′	OC
5.7 m	⊕ 35′	15	IV 2 p	Moderate

This cluster can be seen in binoculars as a faint and small open cluster but is not really impressive in small aperture telescopes. Large apertures show a dozen or so 7th–10th magnitude stars in a rather loose group. Some observers see a Y-shaped asterism on the clusters eastern edge.

Bochum 1	–	06ʰ 25.5ᵐ	+19° 46′	OC
7.9 m	⊕ –	30	I–	Difficult

Now prepare yourself for an observing challenge. This open cluster is only really visible in larger aperture telescopes, where even then it will appear as a very faint collection of stars set against a faint haze of unresolved stars. However, what makes this cluster special is the research done on it that suggests star formation in clusters may be influenced if clusters occur in pairs. A very interesting idea.

NGC 2266	Herschel VI-21	06ʰ 43.3ᵐ	+26° 58′	OC
9.5 m	⊕ 6′	50	II 2 m	Difficult

This is a pleasant cluster though difficult to locate, consisting of over 50 stars tightly compressed. It is believed that the cluster is the youngest that lies so far from the Milky Way's central plane, as other clusters at this distance seem to be at least twice as old. The reason for this difference? As yet no one knows.

NGC 2304	Herschel VI-2	06ʰ 55.2ᵐ	+17° 9′	OC
10.0 m	⊕ 5′	30	II 1 m	Moderate

This open cluster will seem to be a faint haze of about 20 stars when viewed with a moderate aperture, whereas in larger apertures it will appear as a group of 13th magnitude stars set against a larger haze of unresolved stars. It does however have a definite curved aspect to it – its only redeeming feature.

NGC 2331	Collinder 126	07ʰ 07.2ᵐ	+27° 21′	OC
8.5 m	⊕ 18′	30	IV 1 p	Moderate

A large, well spread out open cluster that that makes it difficult to distinguish from the Milky Way. Some observers see a ring-shaped, other an H-shaped, asterism. Oddly enough this is a cluster that hasn't been studied much, if at all. No estimate of its age or distance has been given.

NGC 2355	Herschel VI-6	07^h 16.9^m	+13° 47′	OC
9.7 m	⊕ 9′	34	II 2 p	Easy

After the somewhat lackluster performance of the above two clusters, it's now time for a pleasant change. This open cluster is a welcome relief. It can be easily glimpsed in a large finder and binoculars, and in small to moderate apertures will appear as an irregular grouping of about 30 faint stars. There is a nice yellow star in its center, and at higher magnifications will show many curving arcs of stars. Some observers report that it is actually best seen in small apertures, of say around 10 cm. Try it and see for yourself if you agree.

NGC 2395	Collinder 144	07^h 27.1^m	+13° 35′	OC
8.0 m	⊕ 12′	30	III 1 p	Moderate

This is another bright open cluster that can be glimpsed in a large finderscope. Seemingly split into two groups, or, as some reports suggest, when using larger aperture, more of an oblong shape. It lies far above the galactic plane at about 1,000 light years.

NGC 2420	Herschel VI-1	07^h 38.5^m	+21° 34′	OC
8.3 m	⊕ 10′	100	I 2 r	Easy

This is a delightful open cluster that really stands up well in a larger telescope. With a small aperture it has a circular fuzzy appearance, somewhat resembling a comet. Using averted vision, maybe 20 or more 11th and 12th magnitude stars appear, surrounding a central area of even fainter members. There is a nice arc of stars surrounding the central region that some observers liken to the constellation Corona Borealis.

Notes

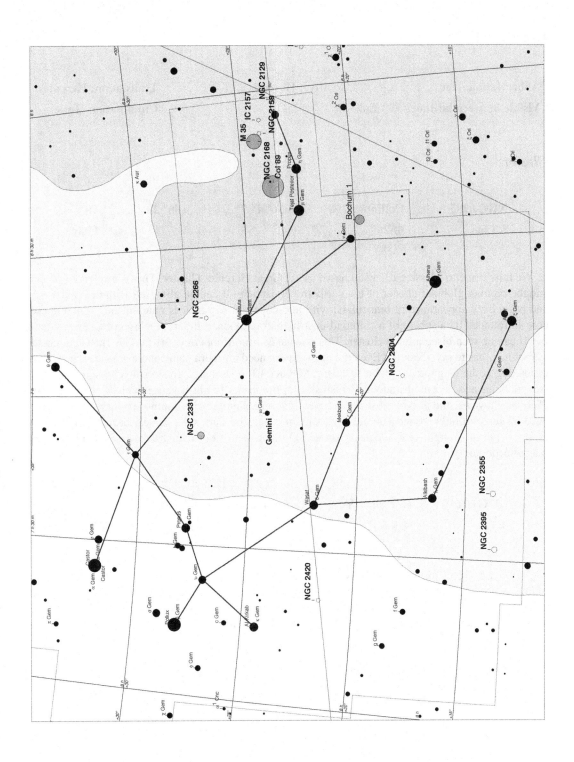

Hercules

Fast Facts

Abbreviation: Her	**Genitive: Herculis**	**Translation: Hercules**
Visible between latitudes 90° and –50°		**Culmination: June**

Clusters

NGC 6205	**Collinder 296**	**16ʰ 41.7ᵐ**	**+36° 28′**	**OC**
5.8 m	⊕ **20′**		**V**	**Easy**

We now come to Messier 13, also known as the Great Hercules Cluster. This is a splendid object and the premier globular cluster of the northern hemisphere. It can be glimpsed with the naked eye and has a hazy appearance in binoculars; with telescopes, however, it is magnificent, with a dense core surrounded by a sphere of a diamond-dust-like array of stars. In larger telescopes, several dark bands can be seen bisecting the cluster. This is what Martin Mobberley, the famous British amateur astronomer, has to say about it: "Even the most experienced northern hemisphere amateurs never tire of looking at the magnificent globular cluster Messier 13 in Hercules. It is a splendid sight, and the best globular at a decent altitude in British skies." It appears bright because is close to us, at only 25,100 light years, and also because it is inherently bright, shining with luminosity equivalent to over 300,000 suns. At only around 160 light years in diameter, the stars must be very crowded, with several stars per cubic light year, a density some 500 times that of our own vicinity. All in all this is a magnificent cluster.

NGC 6229	**Herschel IV-50**	**16ʰ 47.0ᵐ**	**+47° 32′**	**GC**
9.4 m	⊕ **4.5′**		**IV**	**Difficult**

A difficult object to locate, and even with 20 cm telescopes will appear unresolved. It can be glimpsed with binoculars, but of course, it will just appear as a small compact object. Increasing aperture naturally reveals more detail. Large-aperture telescopes will show structure and detail within the cluster. Recent research reveals that it contains a lot of blue stragglers.

NGC 6341	**Messier 92**	**17ʰ 17.1ᵐ**	**+43° 08′**	**GC**
6.3 m	⊕ **14.4**		**IV**	**Moderate**

A beautiful cluster often overshadowed by its more illustrious neighbor, Messier 13. It is a some-what difficult object to locate, but once found is truly spectacular. It can be glimpsed with the naked eye. In binoculars it will appear as a hazy small patch, but in 20 cm telescopes its true beauty becomes apparent with a bright, strongly concentrated core. It also has several very distinct dark lanes running across the face of the cluster. It is a very old cluster, possibly 13 billion years, 25,000 light years distant, and is an Oosterhoff type II (OoII) globular cluster, signifying that it belongs to the group of metal-poor clusters with longer period RR Lyrae variable stars.

Notes

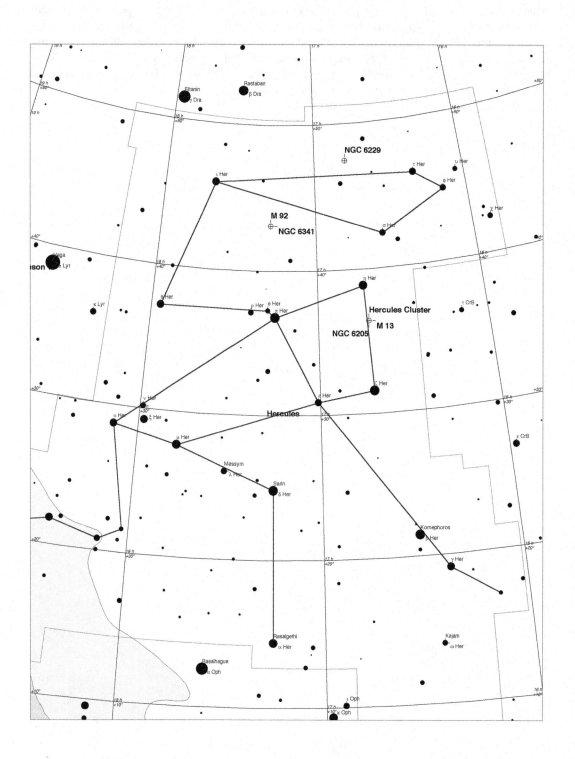

Notes (cont.)

Horologium

Fast Facts

Abbreviation: Hor	Genitive: Horologii	Translation: The Clock
Visible between latitudes 20° and −90°		Culmination: November

Cluster

NGC 1261	Caldwell 87	03ʰ 12.3ᵐ	−55° 13′	GC
8.3 m	⊕ 6.8′	80	II	Moderate

In this small, relatively unobserved constellation is a small, relatively unobserved globular cluster – NGC 1261. It is a small and rather faint object that can prove to be a problem to find and observe, at least in binoculars and small aperture telescopes. What is striking however is that observers report a faint yet definite orangish hue to the cluster. Increasing aperture and magnification will, surprisingly, begin to resolve its inner core, although much will remain granular in appearance. Averted vision will hint at dark rifts and streams of stars, on the border of resolution. Those observers blessed with giant equipment may get a glimpse of the galaxy pair, ESO 155–10 and 10a, about 4. 5′ to the southwest of the cluster, as a tiny nebula-like spot, some 15″ in size.

Notes

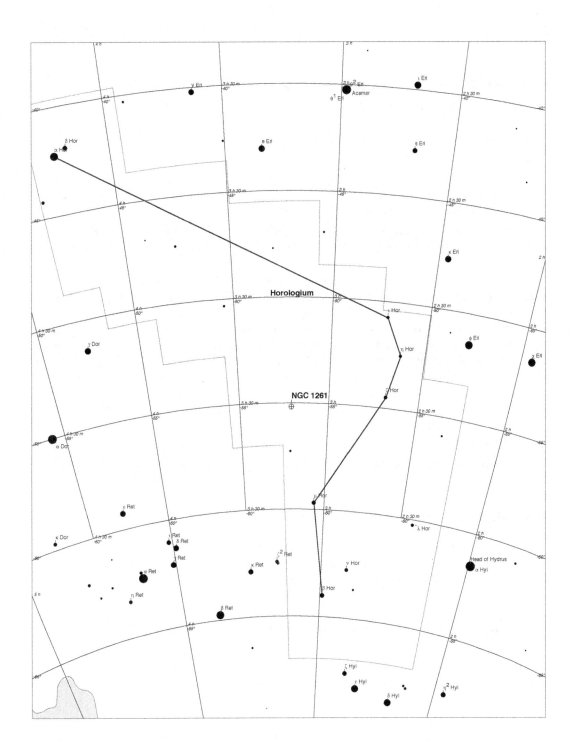

Hydra

Fast Facts

Abbreviation: Hya	**Genitive: Hydrae**	**Translation: The Water Serpent**
Visible between latitudes 60° and −90°		**Culmination: March**

Clusters

NGC 2548	Collinder 179	08ʰ 13.7ᵐ	−05° 45′	OC
5.5 m	⊕ 54′	80	I 3 r	Easy

Located in a rather empty part of the constellation Hydra, this open cluster, also known as Messier 48, is believed to be the missing Messier object. Reported as a naked-eye object, it is nevertheless a nice object in both binoculars and small telescopes. In the former, about a dozen stars are seen, with a pleasing triangular asterism at its center, while the latter will show a rather nice but large group of about 50 stars. Many amateurs find the cluster difficult to locate for the reason mentioned above, but also for the fact that within a few degrees of Messier 48 is another nameless, but brighter, cluster of stars that is often mistakenly identified as Messier 48. Some observers claim that this nameless group of stars is in fact the correct missing Messier object, and not the one that now bears the name.

NGC 4590	Messier 68	12ʰ 39.5ᵐ	−26° 45′	GC
7.3 m	⊕ 11′		X	Easy

Appearing only as a small, hazy patch in binoculars, this is a nice cluster in telescopes, with an uneven core and faint halo. Under low magnification, some faint structure or mottling can be glimpsed that under medium to high magnification resolves itself as a myriad assembly of stars. This is a definite challenge to naked-eye observers, where perfect seeing will be needed. Use averted vision and make sure that your eyes are well and truly dark-adapted.

NGC 5694	Caldwell 66	14ʰ 39.6ᵐ	−26° 32′	GC
10.2 m	⊕ 4.3′		VII	Difficult

This is a faint cluster that has a bright core but an unresolved halo when seen in telescopes of less than 30 cm. An unremarkable object, which you will probably not visit more than once! It is a difficult cluster to locate, especially from the UK. Precise setting circles on your telescope (or of course even a computerized system) will help significantly in finding it. It is actually located on the far side of our galaxy, at around 110,000 light years from the Solar System. Research suggests that the cluster may have attained a velocity that will allow it to escape from the gravitational pull of the galaxy.

Notes

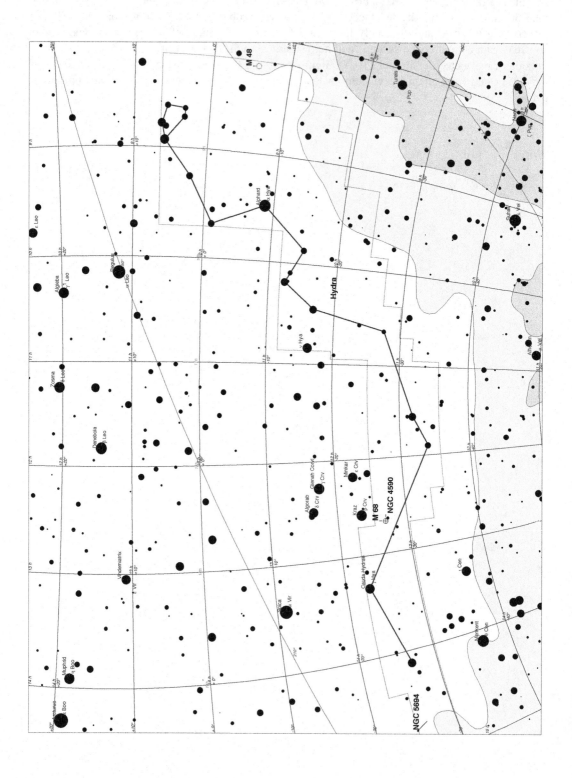

Notes (cont.)

Hydrus

Fast Facts

Abbreviation: Hyi	**Genitive: Hydri**	**Translation: The Water Snake**
Visible between latitudes 5° and −90°		**Culmination: October**

Clusters

NGC 643	–	1ʰ 35.0 ᵐ	−75° 33′	GC
13.0 m	⊕ 0.5′		–	Difficult

Lying in a very sparse constellation, containing very few deep sky objects, is the difficult globular NGC 643. It is small, faint, round, and unresolved except in very large aperture telescopes. It is believed to be an outlying cluster to the Small Magellanic Cloud. Its inclusion here is warranted only as a test of your telescope optics and observing skill.

NGC 1466	–	03ʰ 44.5ᵐ	−71° 40′	GC
11.4 m	⊕ 1.9′		–	Difficult

Another faint, small, unresolved and difficult globular for you to seek out, observe and then move on. Enough said.

Notes

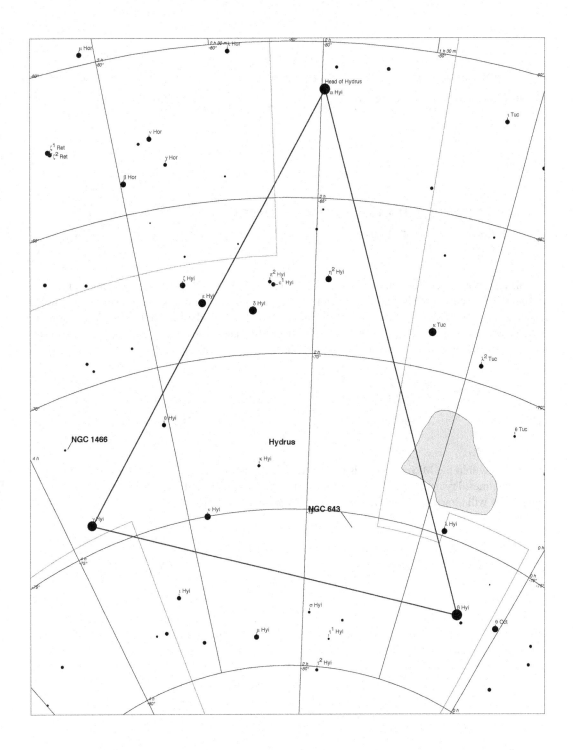

Lacerta

Fast Facts

Abbreviation: Lac	Genitive: Lacertae	Translation: The Lizard
Visible between latitudes 90° and −35°		Culmination: August

Clusters

NGC 7209	Herschel VII-53	22^h 05.1^m	+46° 29′	OC
7.7 m	⊕ 15.0′	25	III 1 p	Easy

This cluster in Lacerta is a fairly bright and irregular open cluster. With the right conditions it can be glimpsed in binoculars and is a fine sight in all telescopes. There is a caveat to this however. The cluster lies in a very rich part of the sky, so that one can be fooled that all the stars in the eyepiece belong to the cluster; they don't. Nevertheless about 100 stars 7th magnitude and fainter can be seen. With increasing aperture lots of star chains and multiple stars will be seen, especially in its eastern sector.

IC 1434	Collinder 445	22^h 10.5^m	+52° 50′	OC
9.0 m	⊕ 7′	40	II 1 p	Easy

Located within the Milky Way, this is a large but irregular cluster of over 70 stars of 10th magnitude and fainter. Try using a high magnification of, say, 150–200X, and also use averted vision. These two factors will almost certainly improve this cluster's observability.

NGC 7243	Herschel VIII-75	22^h 15.0^m	+49° 54′	OC
6.4 m	⊕ 30′	40	IV 2 p	Easy

Also known as Caldwell 16, this cluster is set against the backdrop of the Milky Way. Despite this, this large, irregular cluster nevertheless stands out quite well. Several of the stars are visible in binoculars, but the remainder blur in the background star field. As you would expect, an increase of aperture will show many more members of the cluster. A nice object in an otherwise empty part of the sky – if you overlook the fact that it is located within the Milky Way!

NGC 7245	Herschel VIII-29	22^h 15.2^m	+ 54° 20′	OC
9.2 m	⊕ 6.5′	20	II 1 m	Easy

When using a moderate aperture, around 15 or so stars, aligned north to south, can be seen. It becomes quite a nice, rich cluster at high magnifications and a fine double star will be noticeable at its northern boundary. Increasing the aperture will reveal around 40 stars, at about 13th magnitude and fainter. Under good conditions, you may see a barely tangible background of stars on the threshold of vision. Some observers have reported a large dust lane that appears to pass through the east side of the cluster.

IC 1442	–	22ʰ 16.5ᵐ	+54° 03′	OC
9.1 m	⊕ 3.5′	20	II 2 m	Moderate

Somewhat of a challenge, this open cluster is small and faint consisting of about thirty 11th and 12th magnitude stars. As usual with such dim objects, the larger the aperture, the better.

NGC 7296	Herschel VII-41	22ʰ 28.0ᵐ	+52° 19′	OC
9.7 m	⊕ 3.0′	20	III 2 p	Moderate

Although this open cluster is faint and small, it is fairly easy to recognize. However, it is best to try to observe it with moderate aperture and larger, along with moderate magnification, although that being said, it can nevertheless be seen with small instruments. There are about 25 or more stars in the cluster, ranging in magnitude from 10th to 13th.

Notes

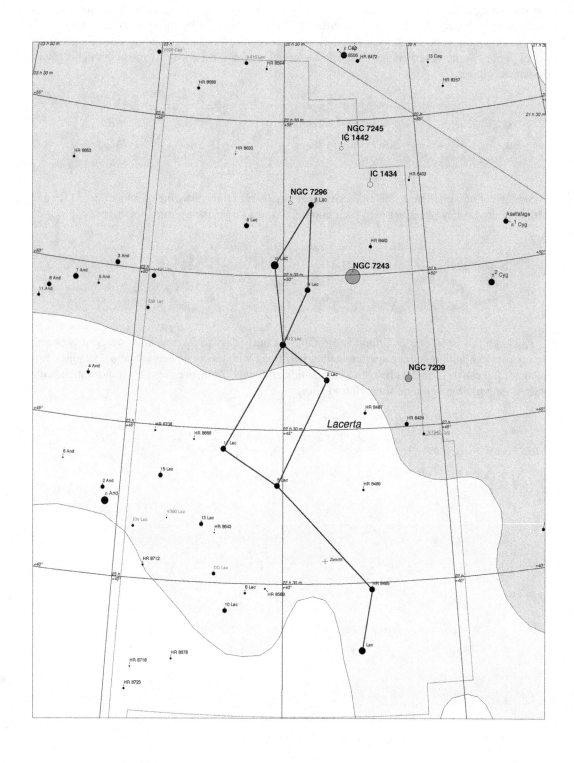

Notes (cont.)

Lepus

Fast Facts

Abbreviation: Lep	**Genitive: Leporis**	**Translation: The Hare**
Visible between latitudes 60° and −90°		**Culmination: December**

Clusters

NGC 1904	Messier 79	05ʰ 24.4ᵐ	−24° 32′	OC
8.6 m	⊕ 8.7′		V	Moderate

A fine open cluster, but best appreciated with a telescope because binoculars cannot resolve it. Small telescopes of, say, 10 cm aperture can resolve the core, but it will be a challenge because you will need a high magnification, dark skies and averted vision. Therefore this is a perfect test for you and your telescope optics. Telescopes of aperture 40 cm and more will resolve the core with no difficulty. Located at a distance of 41,000 light years, it is the sole globular cluster for northern hemisphere observers in the winter. Like Messier 54, it is believed that Messier 79 is in fact a member of the Canis Major dwarf galaxy, currently undergoing a close encounter with the Milky Way, during which it will eventually be eaten in the act of galactic cannibalism (a real term!).

NGC 2017	–	05ʰ 39.4ᵐ	−17° 51′	OC
−m	⊕ 3.3′ × 2.4′		II 1 p	Open/Double?

Now for something completely different; this is either a very small and poorly represented open cluster, an asterism, or it is a very impressive multiple star system! It goes something like this. The "cluster" has five bright members and four fainter. In addition, four of the brighter stars constitute two doubles stars. To split either of these will require high magnification and large apertures, but they do show nice color contrasts of yellow and orange (to some eyes anyway). There are some reports that it can be glimpsed in binoculars, but can be seen in apertures of 10 cm and greater. Definitely a cluster to look out for.

Notes

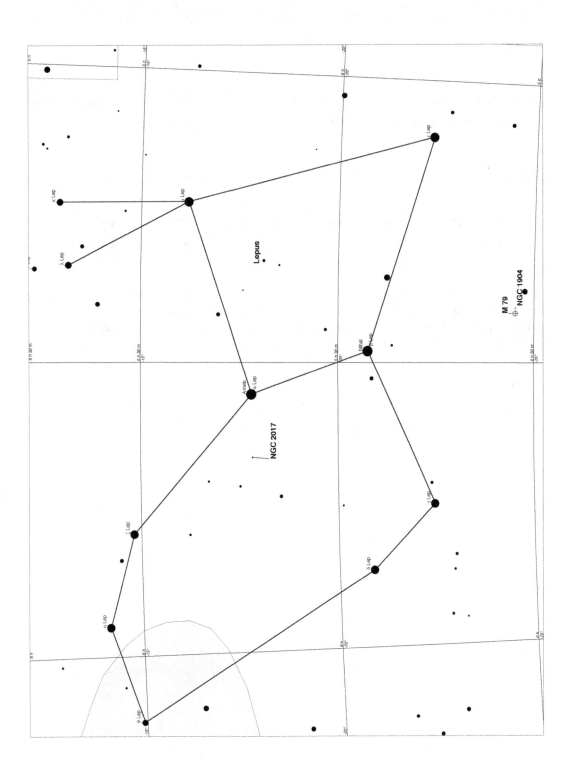

Libra

Fast Facts

Abbreviation: Lib	Genitive: Librae	Translation: The Beam Balance
Visible between latitudes 65° and −90°		Culmination: May

Clusters

NGC 5897	Herschel VI-19	15ʰ 17.4ᵐ	−21° 01′	OC
8.4 m	⊕ 13′		XI	Difficult

This is a very difficult cluster to locate even in binoculars, owing to its low surface brightness. With a telescope of medium aperture, say, 20 cm, and with averted vision some dozen stars may be glimpsed. But with apertures of 40 cm or more, the cluster will be resolved and show a much larger aspect.

Notes

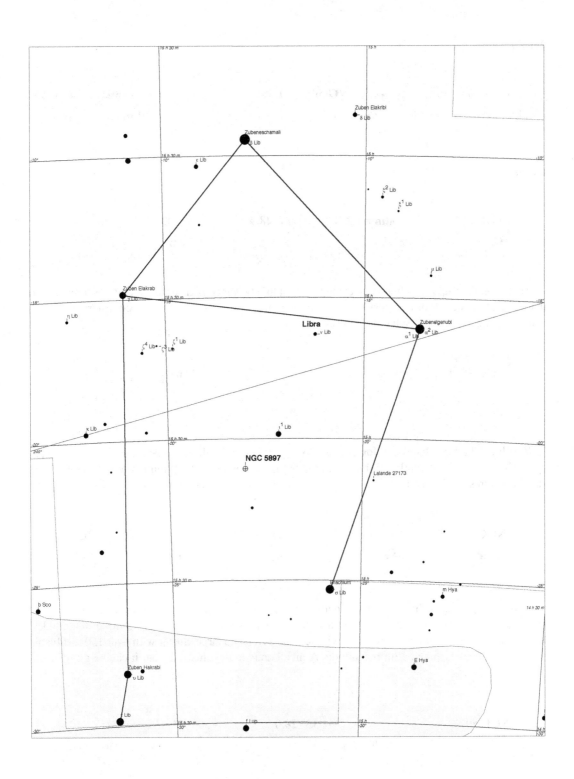

Lupus

Fast Facts

Abbreviation: Lup	Genitive: Lupi	Translation: The Wolf
Visible between latitudes 35° and −90°		Culmination: May

Clusters

NGC 5749	Collinder 287	14^h 48.9^m	−54° 030	OC
8.8 m	⊕ 7′	20	II 2 m	Difficult

This object in Lupus is a fine open cluster set in a lovely star background. Several observers report that there at the center of the cluster is a suggestion of a wishbone, or "Y"-shaped, asterism of 10th and 11th magnitude stars. What do you see?

NGC 5824	–	15^h 04.0^m	−33° 04′	GC
6.2 m	⊕ 9.1′		I	Difficult

You'll need a large telescope for this one. And even then it will just appear as a small but bright spot. The core is significantly brighter than the halo, but the cluster will remain unresolvable. With small apertures it will just look like a tiny object with a barely discernible hint of a brighter core.

NGC 5822	Collinder 289	15^h 04.3^m	−54° 23′	OC
6.5 m	⊕ 39′	60	II 2 r	Easy

A very large cluster that will fill the entire field of small and medium aperture telescopes at low magnification. It is quite well spread out and will appear as if all the stars have similar magnitudes. But it really excels at high aperture, as it will become a rich and large cluster with over 150 resolvable stars, all at about 11th and 12th magnitude. A nice bonus to the cluster is that it can be glimpsed in binoculars.

NGC 5927	–	15^h 28.0^m	−50° 470	GC
8.0 m	⊕ 6.0′	8		Moderate

This globular cluster is best appreciated with a large aperture, when it will appear as a bright, large and circular object. There is a slight brightening of the core, and around a 20 or so stars can be resolved in the halo. With an increase of magnification, many more halo stars become visible, and there is always the hint that more are just waiting to be resolved. At medium apertures, the cluster will appear small, although bright, and no resolution will be achieved.

NGC 5986	–	15^h 46.1^m	–37° 47′	GC
7.8 m	⊕ 9.6′		VII	Easy

In an oft-passed over constellation, usually because only some of it is visible from Europe and the United States, we find a nice globular cluster that can be seen in binoculars. In small apertures it will show a fairly bright and condensed core enveloped by fainter 13th magnitude stars. Some observers report it has a faint, nearly intangible color. Some say it is bluish, while others yellowish. It may be dependent on the magnification and aperture one uses. If you have the opportunity or means to see this, then do so, as it is a nice object.

Notes

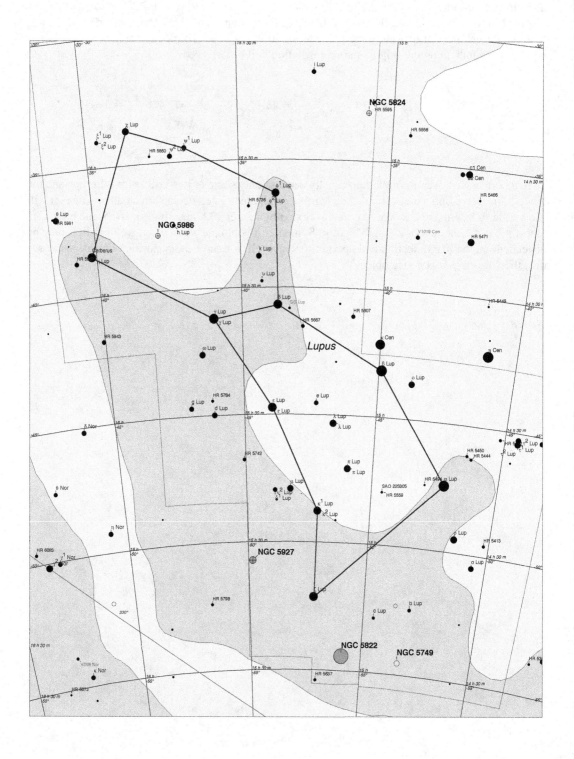

Notes (cont.)

Lynx

Fast Facts

Abbreviation: Lyn	Genitive: Lyncis	Translation: The Lynx
Visible between latitudes 90° and −35°		Culmination: January

NGC 2419	Caldwell 25	07^h 38.1^m	+38° 53′	GC
10.4 m	⊕ 4.6′		II	Difficult

This globular cluster, also known as the Intergalactic Wanderer, can be a difficult object to resolve with any detail, when even large telescopes of aperture 40 cm will be unable to resolve any stellar detail. It has been reported as observed with a 10 cm telescope under perfect conditions, but will appear as just a faint, hazy dot. It lies at a distance of 300,000 light years, further even than the Magellanic Clouds, because it has a space velocity in excess of the velocity needed to escape from the gravitational pull of our galaxy. One of the five most luminous globular clusters to be found in the Milky Way, research has suggested that it is in fact of extragalactic origin and maybe the remains of a dwarf galaxy, captured and subsequently disrupted long ago by the Milky Way.

Notes

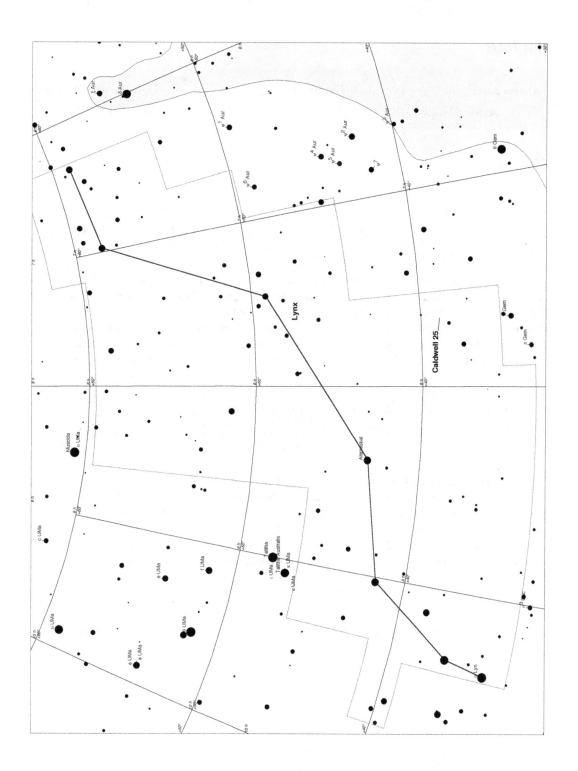

Lyra

Fast Facts

Abbreviation: Lyr	Genitive: Lyrae	Translation: The Lyre
Visible between latitudes 90° and −40°		Culmination: July

Clusters

Stephenson 1	–	18^h 53.5^m	+36° 55′	OC
3.8 m	⊕ 20′	40	IV 3 p	Easy

Often overlooked, this open cluster is a delight for several reasons. It can be seen in both binoculars and small telescopes and will present a fairly large but loose group of stars, lying between δ^1 (11) and δ^2 (12) Lyrae. Both these stars incidentally are thought to be true members of the cluster. In addition, these two are a lovely colored double, with δ^2, magnitude 4.5, an orange M4 giant, and δ^1, magnitude 5.5, a blue-white B3 main sequence star. Increasing aperture and magnification shows many more cluster members. Current studies indicate that the cluster is very close to us at 800 light years.

NGC 6779	Messier 56	19^h 16.6^m	+30° 11′	GC
8.3 m	⊕ 8.8′	–	X	Moderate

This pleasing globular cluster is situated among a rich star field and in small aperture will appear as a hazy patch with a brighter core. It can be a challenge to locate if using binoculars. It has often been likened to a comet in its appearance. Resolution of the cluster will need at least a 20 cm aperture telescope, and increasing magnification will show further detail. Recent research suggests that the cluster may have been formed during the merger of a cannibalized dwarf galaxy, the surviving nucleus of which forms the famous globular cluster Omega Centauri (see entry in constellation Centaurus).

NGC 6791	–	19^h 20.7^m	+37° 46′	OC
9.5 m	⊕ 16′	300	I 2 r	Moderate

This is a pleasant rich open luster of faint stars. It contains many faint 11th-magnitude stars and so poses an observing challenge for medium apertures, but in larger telescopes with high magnifications, literally hundreds of barely resolved stars could be seen. It is one of the oldest globular clusters in the sky, and thus metal-rich, and as a consequence it is also one of the most studied. This is an object especially suited for large aperture telescopes.

Notes

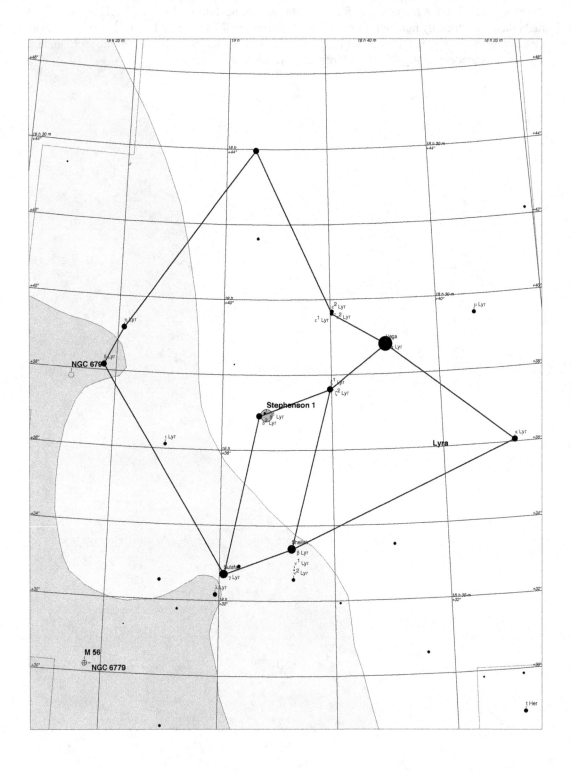

Notes (cont.)

Mensa

Fast Facts

Abbreviation: Men	Genitive: Mensae	Translation: The Table Mountain
Visible between latitudes 0° and –90°		Culmination: December

Cluster

NGC 1848	–	05ʰ 07.4ᵐ	–71° 11′	OC/Ast/?
9.7 m	⊕ 3′		–	Moderate

Mensa's only open cluster that is to be included here (there are several even fainter ones that will not be mentioned because they need very large telescopes) is buried deep within a nebula, a favorite for astrophotography (the nebula, not the cluster). It is a scattered and loose group or less than a dozen stars. The question is, is it a true cluster or an asterism?

Notes

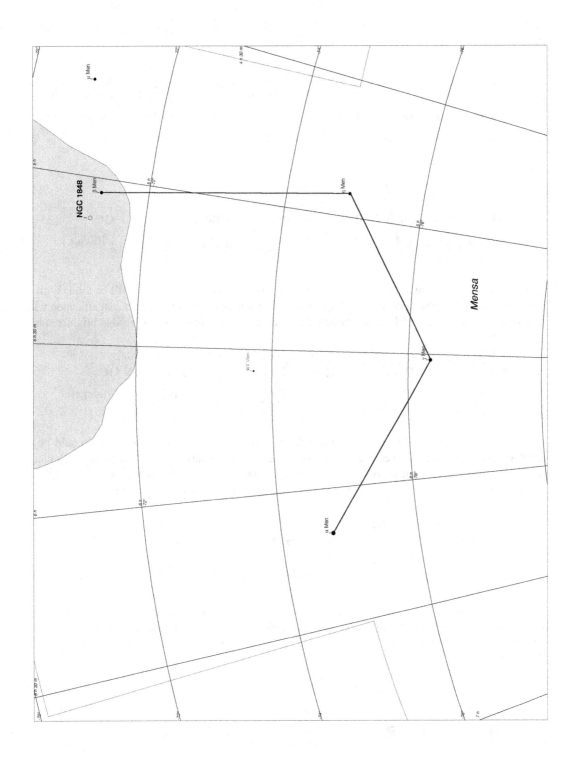

Monoceros

Fast Facts

Abbreviation: Mon	Genitive: Monocerotis	Translation: The Unicorn
Visible between latitudes 75° and −85°		Culmination: January

Clusters

NGC 2269	Collinder 114	06^h 00.2^m	−10° 37′	OC
10.0 m	⊕ 4′	12	II 1 p	Difficult

The first things that must be said is that even in a large telescope, this cluster will be faint. With a low magnification, it will appear as a faint haze set against a nice star field. Higher magnification will resolve about 20–30 stars of 13th magnitude. At the center of the cluster is a somewhat brighter star.

NGC 2215	Herschel VII-20	06^h 20.8^m	−07° 17′	OC
8.4 m	⊕ 8.0′	40	II 2 p	Moderate

Small and faint is an adequate description for this cluster, and in an urban location could be a problem, even though it can be seen in binoculars. However, paradoxically, it becomes much nicer in telescopes when, in small apertures, four 10th–11th magnitude stars can be seen in the core of the cluster, whereas increasing aperture reveals rows of stars and scattered fainter members.

Collinder 91	–	06^h 21.7^m	+02° 21′	OC
6.4 p	⊕ 17′	20	II 2 p	Easy

Many clusters exhibit a definite shape when seen in the eyepiece, and this is no exception. Observers have reported it as a diamond-like group of stars; sadly however, it is so meager that it can be difficult to differentiate from the background star field. Some people regard this as nothing more than an asterism.

Collinder 92	–	06^h 22.9^m	+05° 07′	OC/Ast/?
8.6 m	⊕ 11′	–	–	–

Is this really a cluster? Maybe not. There is a small and faint group of less than a dozen stars, but should we call it a cluster or an asterism?

NGC 2232	Herschel VIII-25	06^h 27.2^m	−04° 45′	OC
4.2 m	⊕ 53′	20	IV 3 p	Easy

A large, irregular cluster that is associated with the 5th magnitude star 10 Monocerotis. It's fairly bright, fortunately, that allows its brightest members to be seen, with binoculars, even from a heavily light-polluted location. High magnification, and the cluster now appears richer and has a more well formed aspect.

NGC 2236	Collinder 94	06^h 29.6^m	−06° 50′	OC
8.5 m	⊕ 6′	50	II 2 m	Moderate

This cluster can be seen in large binoculars, and with medium aperture will appear as a nice, round haze embedded in which will be a few resolved 11th and 12th magnitude stars. As to be expected, an increase in aperture will reveal more of its fainter members.

Collinder 96	–	06^h 30.3^m	+02° 52′	OC
7.3 m	⊕ 7′	15	IV 2 9	Easy

Now for a cluster that is easy to locate. Just search for the nice 6th magnitude double star Bail 1695, that has a separation of roughly 20¢, then follow the trail of 9th magnitude stars that lead to the northeast, and there is a conspicuous group of stars arranged in a semicircle. You have found the cluster.

Collinder 97	–	06^h 31.3^m	+05° 55′	OC
5.4 p	⊕ 21′	15	IV 3 p	Easy

Another large and scatter cluster. In moderate apertures it will be seen the cluster is enclosed by a triangle of stars, all around the 6th magnitude. There are many more 9th magnitude and fainter stars in the group that become apparent with increasing aperture and magnification.

NGC 2250	Collinder 100	06^h 34.1^m	−05° 00′	OC
8.9 m	⊕ 7′	10	IV 2 p	Difficult

Many observing guides list this as a cluster, consisting of just a few 12–14th magnitude stars, but some think it looks more like a random group of stars. Maybe using a very large aperture will show its true nature.

NGC 2251	Herschel VIII-3	06^h 34.6^m	+08° 22′	OC
7.3 m	⊕ 10.0′	30	III 2 p	Difficult

Another cluster that can be a problem to see from an urban location, so dark skies will be needed, even for binoculars. In any telescope it will be distinct, although larger aperture will reveal the fainter members, with a quite a few red-orange stars. What can be said is, however, that with averted vision and perhaps high magnification, it may appear to have a sinuous shape? In its southern area a slight nebulous haze to the field of view may be glimpsed, under the right conditions, of course.

NGC 2252	Collinder 102	06^h 34.7^m	+05° 22′	OC
7.7 m	⊕ 20′	20	III 2 m n	Moderate

Although this can be seen as a scattered, irregularly shaped cluster of about 30 or more stars, with magnitudes ranging from 10 to 12.5, what makes it special is that it lies on the edge of the famous Rosette Nebula.

NGC 2254	Collinder 103	06^h 35.9^m	+07° 40′	OC
9.1 m	⊕ 4′	30	I 1 m	–

In a small aperture the cluster will be small and faint with about six stars visible, and an unresolved, barely acknowledged haze of background stars. An increase in aperture shows more of its fainter members. Not one of the constellations best efforts.

Collinder 104	–	06^h 36.5^m	+4° 50′	OC
9.6 m	⊕ 21′	15	IV 2 p n	Moderate

Lying to the west of Collinder107 is this much fainter cluster. It is easy to observe as a long chain of stars, about 20′ long, in a north to south direction. Best seen in moderate to large apertures.

Collinder 106	–	06^h 37.1^m	+05° 57′	OC
4.6 p	⊕ 45′	20	III 3 p	Easy

This is a large, scattered group of stars, several of which range from 6th to 7th magnitude. This means that, theoretically, the cluster may be visible to the naked eye, or at least its brightest members. There seem to be no reports of it being seen.

Collinder 107	–	06ʰ 37.7ᵐ	+04° 44′	OC
5.1 p	⊕ –′	16	IV 3 p	Easy

Here again we have a cluster whose brightest stars are within naked-eye range. Admittedly, all the other stars are beyond the range, with the cluster proper well scattered, large and bright, but nevertheless it is questionable if anyone has ever made an attempt to see the cluster, or its brightest stars.

Collinder 111	–	06ʰ 38.7ᵐ	+06° 54′	OC/Ast
7.0 m	⊕3.2′	–	–	–

This object doesn't really deserve the classification of a cluster, as it seems to me to be just a meager collection of very faint stars. Observe and decide for yourself.

NGC 2264	Herschel VIII-5	06ʰ 41.0ᵐ	+09° 54′	OC
4.4 m	⊕ 40.0′	40	IV 3 p	Easy

This is a splendid naked-eye object (under perfect conditions), and even better in binoculars, where it will show about 15 or more stars, in the now-familiar shape of a Christmas tree, hence its popular name, The Christmas Tree cluster. It gets even better when using any sized telescope; using moderate and larger apertures it will show the faint nebula associated with the cluster, and under very dark skies, and a large telescope, the Cone Nebula may be glimpsed surrounding a 7th magnitude star at the top of the tree, although this would be a challenge to see from an urban setting.

Collinder 115	–	06ʰ 46.5ᵐ	+01° 46′	OC
9.2 p	⊕ –7	50	II 2 p	Difficult

Another one of the faint clusters that Monoceros has in plenty, Collinder 115 is a group of about fifty 12th magnitude and fainter stars. It can only be seen as a distinct cluster in large apertures, and even then moderate magnification will be useful; otherwise, low magnification and aperture will just show it as a coarse-textured haze.

NGC 2286	Herschel VIII-31	06ʰ 47.7ᵐ	–03° 09′	OC
7.5 m	⊕ 15′	50	IV 3 m	Moderate

Although the cluster is fairly large, it is nevertheless faint, and to add to its trials it is located in a fairly obscure part of the Milky Way. Assuming you live where dark skies are the norm, it will be visible in binoculars. However, any light pollution will make it a problem to locate. Telescopically, it is difficult to make out the cluster from the background, and it is really more suited for larger apertures, where the faint stars can be resolved.

NGC 2301	Herschel VI-27	06^h 51.8^m	+00° 28′	OC
6.0 m	⊕ 15′	70	I 3 m	Easy

In a word – superb! Here we have a very striking cluster. In binoculars, a north–south chain of 8th and 9th magnitude stars is revealed, marked at its midway point by a faint haze of unresolvable stars. With large aperture, there is a colorful trio of red, gold and blue stars at the cluster's center. Also, you cannot but fail to see the rich background of the Milky Way in any field of view, because the cluster lies approximately on the Galaxy's mid plane. It can be said, with some justification, that this may be the constellation's finest cluster.

NGC 2302	Herschel VIII-39	06^h 51.9^m	−07° 05′	OC
8.9 m	⊕ 2.5′	30	II 2 p	Moderate

A moderate to large aperture will be needed for this cluster, when it will appear a fairly small group of about two dozen stars embedded in a large, scattered field of stars. On its western side is a nice group, including three 10th magnitude stars. On its eastern side is a nice double and a triple star forming either a "V" or "Y" asterism, depending on what you see. As an aside, it is believed that NGC 2299 is probably the same cluster as NGC 2302, and thus is one of the "nonexistent" objects.

NGC 2323	Collinder 124	07^h 03.2^m	−08° 20′	OC
5.9 m	⊕ 16′	80	II 3 m	Easy

The only Messier object in Monoceros, Messier 50 is one that is often overlooked by amateurs. Discovered by Cassini, this is a fine, heart-shaped cluster easily seen in binoculars, and visible to the naked eye on clear nights. Within the large, bright and irregular cluster of blue stars is a striking red star. What makes the cluster particularly challenging is that the area of the sky where it resides is full of small stellar groupings and asterisms. The question often arises, where does the cluster end and the background star fields begin?

NGC 2324	Herschel VII-38	07^h 04.1^m	+01° 03′	OC
8.4 m	⊕ 8.0′	70	II 2 r	Moderate

This cluster is best seen with a moderate aperture telescope, and only then it will appear as faint and small. If you are using a small aperture, then averted vision will be needed here. Higher magnification reveals a uniform haze of faint starlight. Naturally, a large aperture will reveal many more stars.

NGC 2335	Herschel VIII-32	07^h 06.8^m	−10° 02′	OC
7.2 m	⊕ 7.0′	35	III 3 m n	Moderate

This open cluster can pose a problem to find under urban skies, as the light pollution may prevent you from catching this one. It is small, although fairly bright, but not particularly special. With a low magnification, observers have reported that it has a definite "Y" shape. Using higher magnification and aperture will allow many more faint stars to be glimpsed.

NGC 2343	Herschel VIII-33	07^h 08.1^m	−10° 37′	OC
6.7 m	⊕ 6.0′	20	III 3 m n	Easy

Visible in binoculars, this is a rich cluster, somewhat bright, and its appearance only improves when telescopes. Small telescopes will reveal a handful of fairly bright stars, and naturally use of a higher magnification will reveal its fainter members. Larger telescopes will allow you to see the fainter stars in the group.

NGC 2353	Herschel VIII-34	07^h 14.5^m	−10° 16′	OC
7.1 m	⊕ 18′	30	II 2 p	Easy

This is a bright and tight cluster that can be glimpsed in binoculars, under a dark sky, as a 6th magnitude, slightly orangish star enclosed within what looks like a nebula, but is in fact, unresolved stars. Two slight arcs of stars can be seen, and with higher magnification many fainter 11th and higher magnitude stars can be seen, maybe as many as 70–80. It is interesting to note that if the 6th magnitude star, mentioned earlier, is seen with the naked eye, then surely that would mean that the cluster is visible to the naked eye, or is that taking the definition of a naked eye object just a bit too far? You decide!

NGC 2506	Collinder 170	08^h 00.2^m	−10° 47′	OC
7.6 m	⊕ 8′	100	I 2 r	Moderate

A nice rich and concentrated cluster best seen with a telescope, but one that is often overlooked owing to its faintness even though it is just visible in binoculars. Caldwell 54 includes many 11th and 12th magnitude stars. It is a very old cluster, around 2 billion years, and contains several blue stragglers. These are old stars that nevertheless have the spectrum signatures of young stars. This paradox was solved when research indicated that the young-looking stars are the result of a merger of two old stars.

Notes

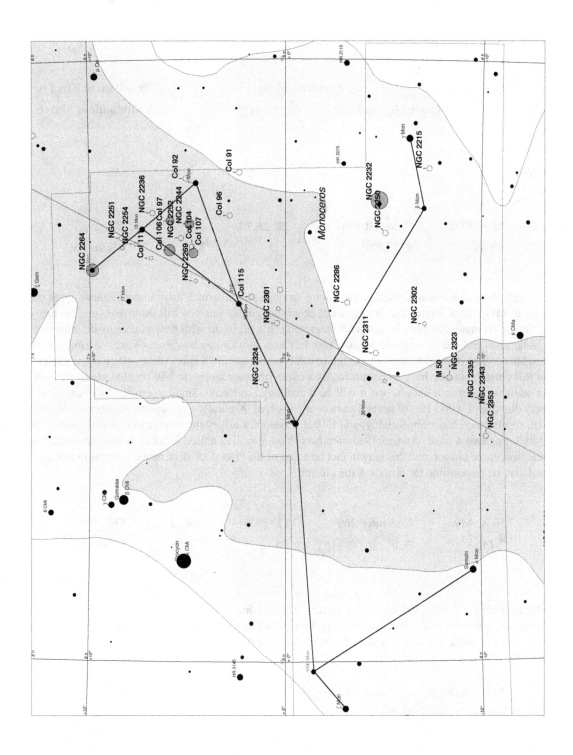

Musca

Fast Facts

Abbreviation: Mus	**Genitive: Muscae**	**Translation: The Fly**
Visible between latitudes 10° and −90°		**Culmination: March**

Clusters

NGC 4372	**Caldwell 108**	**12ʰ 25.7ᵐ**	**−72° 40′**	**GC**
6.9 m	⊕ **13.5′**		**XII**	**Moderate**

Every once and a while, there comes along an object about which astronomers cannot agree on. This is one of them. Even a cursory glance at observing guides and lists will show that there are many differing estimates for both its size and magnitude. It's all to do with how you estimate where the cluster ends and the background begins, and thus how to estimate how many stars contribute to its brightness. The values given above are the best estimates (for now). Binoculars will give you glimpse of this cluster, but you have to be careful, as a close 7th magnitude star will provide plenty of glare. In addition, even in a finderscope it will be a challenge to locate. Small apertures will just show a very dim object with a hint of several stars being resolved. Naturally large aperture and magnification will resolve the cluster appreciably, and in fact it becomes a very impressive object with around 50 stars set against a haze of unresolved members. You may also notice the dark lane at its northwest that pierces the cluster, and this may in fact be a part of the "Doodad" dark nebulae that is in this area, and also be responsible for dimming the cluster.

NGC 4463	**Collinder 260**	**12ʰ 29.9ᵐ**	**−64° 47′**	**OC**
7.2 m	⊕ **5′**	**30**	**I 3 m**	**Moderate**

Not only is this a rather nice open cluster, but it is also located in a fabulous part of the sky and also near the edge of the Coalsack dark nebula. About a dozen or so stars can be resolved easily that some say looks like a tree with faint stars radiating out just like branches. Increasing aperture will of course reveal more cluster stars, even though it lies in a rich star field.

NGC 4815	**Collinder 265**	**12ʰ 58.0ᵐ**	**−64° 58′**	**OC**
8.6 m	⊕ **3′**	**50**	**II 2 r**	**Moderate**

This is a small and faint open cluster that lies just within the southern edge of the dark nebula, the Coalsack. Being within the nebula probably accounts for some of its dimness, as the light will suffer extinction. Nevertheless it is an impressive sight in moderate to large apertures and takes magnification well when it will reveal several resolved stars set amongst an unresolved haze. Several of the stars show delicate hues of color, yellow and yellowish-orange in particular. Are these colors indicative of the temperature of the stars, or just the effect of interstellar reddening? No doubt someone will find out eventually.

NGC 4833	Caldwell 105	12ʰ 59.6ᵐ	−70° 52′	GC
8.4 m	⊕ 13.5′		VIII	Easy

Easily located, this often-overlooked globular cluster can be glimpsed in binoculars and even a finderscope, but of course not resolved. In fact it would be a challenge to resolve even in small apertures unless under very dark skies. It has a large and broad core that tends to overwhelm the fainter halo. Using a larger aperture will increase the appeal of the cluster, where definite granularity in the core is revealed, with a plethora of 13th and 14th magnitude stars in the halo. This is a very nice object to observe, and in fact will repay long and detailed study. It actually lies in a very dusty part of the Milky Way, which will decrease the amount of light that we receive.

Collinder 277	–	13ʰ 48.7ᵐ	−66° 05′	OC
9.2 m	⊕ 12′	30	IV 1 m	Moderate

Although this is a relatively faint cluster, it is obvious, and in moderate and large apertures presents a largish and irregular group of about 40 stars, most of them at 11th and 12th magnitude. When using a large aperture many star chains and loops can be seen.

Notes

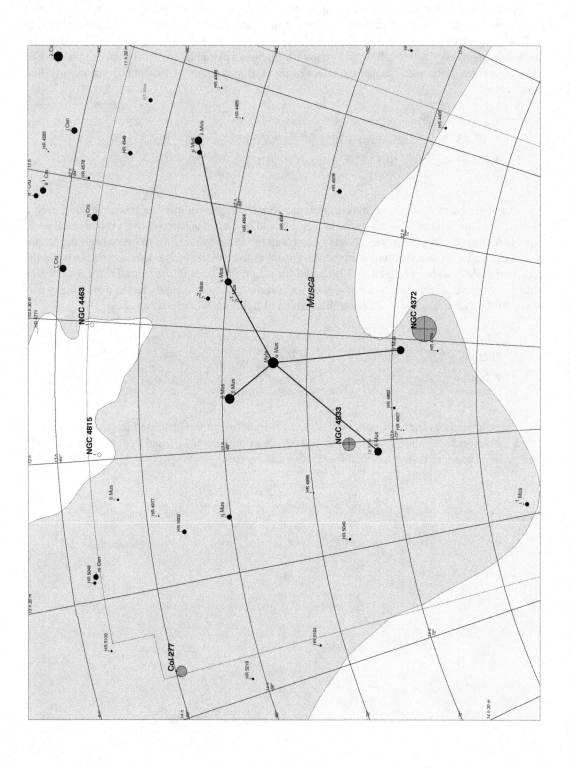

Notes (cont.)

Norma

Abbreviation: Nor	Genitive: Normae	Translation: Carpenters Square
Visible between latitudes 30° and –90°		Culmination: May

NGC 5925	Collinder 291	15^h 27.4^m	–54° 31'	OC
8.4 m	⊕ 20'	50	III 2 r	Difficult

A difficult cluster to see as it merges into the background star fields, and thus it is a problem to say which is cluster and which is not. The use of a medium to large aperture will help somewhat, and when located will consist of about four dozen 11th and 12th magnitude stars scattered over a large area.

NGC 5946	IC 4550	15^h 35.4^m	–50° 40'	GC
8.4 m	⊕ 3'		IX	Difficult

Due to its size and faintness, it will appear, not surprisingly, small and faint even in a moderate aperture, with no resolution of core or halo. Larger aperture and high magnification may resolve a handful of stars at the edge of the halo.

Collinder 292	–	15^h 50.7^m	–57° 40'	OC
7.9 m	⊕ 15'	50	IV 2 m	Easy

After the slightly disappointing previous entries, regarding observability, we now have an open cluster that really shines in binoculars or a wide field telescope, and there are hints that it can even be glimpsed in a finderscope. It may look as if it is nothing more than a distinct enhancement of the Milky Way, but that does not deter from its pleasing aspect. Using a medium or large aperture adds nothing.

NGC 5999	Collinder 293	15^h 52.1^m	–56° 28'	OC
9.0 m	⊕ 3.0'	40	I 2 m	Moderate

A medium to large aperture will be needed here, as this open cluster is small, sparse and fairly faint. Most of the stars are 12th and 13th magnitude, and at highest magnification a void will seem to be at its center, surrounded by a striking circle of stars.

NGC 6031	Collinder 297	16^h 07.6^m	–54° 01'	OC
8.5 m	⊕ 2.0'	120	I 3 p	Easy

An easy to find cluster one degree northwest of the yellow 5th magnitude star Kappa (k) Normae, which is helpful, as this is a small and faint cluster. Even with a large aperture it will look like a faint glowing spot with only a handful of resolved stars. A high magnification will only resolve more 12th and 13th magnitude stars.

NGC 6067	Collinder 298	16ʰ 13.2ᵐ	−54° 13′	OC
5.6 m	⊕ 12′	300	I 3 r	Easy

At last we have a stunning open cluster! This is visible to the naked eye on dark nights and is a wonderful spectacle in all apertures. Often compared to the best of the Messier objects, it is very large and rich and will fill the field of view with literally hundreds of both bright and faint stars. There is a lovely orange star near its center and a very red star to its north. Increasing aperture reveals almost limitless stars, many in arcs, chains, loops, and voids, along with double stars and triple stars. In my opinion a gem of the southern sky.

NGC 6087	Collinder 300	16ʰ 18.9ᵐ	−57° 56′	OC
5.4 m	⊕ 15′	40	II 2 m	Easy

One of the brightest and largest open clusters in Norma. Caldwell 89 is quite an impressive object with a very impressive "V"-shaped asterism. The stars range from 9th to 13th magnitude, and the cluster is visible to the naked eye with the usual caveat, although averted vision will help in its discovery. It takes well to binoculars, with a chain of stars set against an unresolved background. Any telescopic aperture tends to lose the cluster aspect, and shows instead a rich tapestry of loops, arcs and chains, more or less similar to any rich part of the Milky Way. Incidentally, its brightest star is the orange Cepheid variable star S Normae.

NGC 6115	Ruprecht 118	16ʰ 24.4ᵐ	−51° 56′	OC
9.8 m	⊕ 3.4′	7	I 2 p	Moderate

An open cluster that is not often observed, and not without reason, as it is small, faint and doesn't actually contain many stars. Having said that, however, it is immediately obvious in medium to large apertures, with about one dozen stars set in a rich field. Not very impressive, but not without worth either.

NGC 6134	Collinder 303	16ʰ 27.7ᵐ	−49° 09′	OC
7.2 m	⊕ 6.0′	50	II 3 m	Easy

This is a very nice open cluster that actually looks rather good through a finderscope and has a collection of both bright and faint stars. A loose but rich cluster that stands out well against the background and benefits from an increase of both aperture and magnification.

NGC 6152	Collinder 304	16ʰ 32.7ᵐ	−52° 38′	OC
8.1 m	⊕ 29′	70	III 3 r	Easy

A very big cluster, similar in size to the Moon, set against an absolutely wonderful background of the Milky Way, thus making the cluster best viewed at low magnification. There are close to a hundred stars of 10th magnitude and fainter when seen in large apertures, but any telescope will show a pleasing aspect.

NGC 6169	Collinder 306	16ʰ 34.1ᵐ	−44° 03′	OC
6.6 m	⊕ 6′	40(?)	III 1 m	Easy

An often-overlooked object, also called the Mu (μ) Normae cluster. Although fairly bright the view is dominated by the blue supergiant star, after which it is named, and this tends to obscure the cluster stars. To attempt to see the cluster, try a medium or large aperture, and move the telescope so that the star is out of the field of view. Admittedly you can only see one half of the cluster at any one time, but it is better than not seeing it at all. Is this a naked-eye object? There have been few reports as to its validity.

Notes

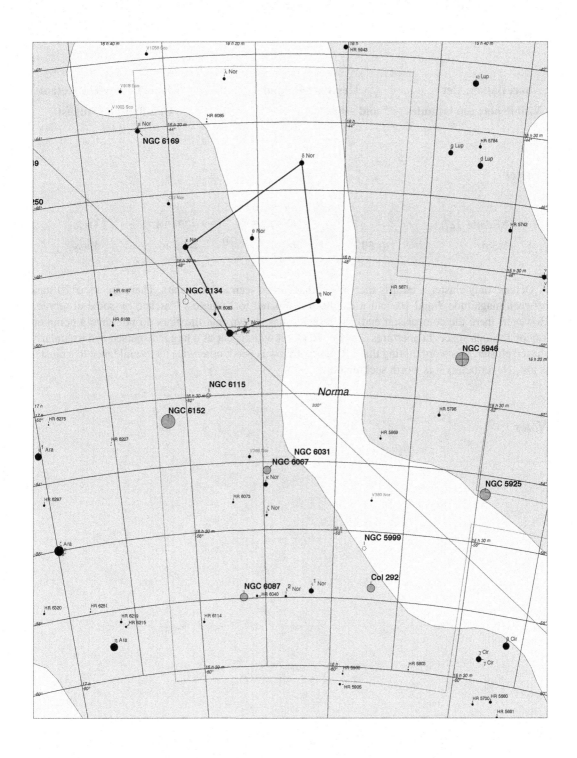

Octans

Fast Facts

Abbreviation: Oct	**Genitive: Octantis**	**Translation: The Octant**
Visible between latitudes −5° and −90°		**Culmination: August**

Cluster

Melotte 227	–	20ʰ 17.3ᵐ	−79° 18′	OC
5.3 m	⊕ 50′	40	III 3 m	Easy

Octans' only cluster is a fairly nice object that can be seen in binoculars. There are about 20 stars between magnitude 7 and 10, and it has been compared to Kemble's Cascade by some observers. However, there the comparison ends, as research has shown that the stars do not have a common motion through space. In apertures of about 25 cm, it will appear as a large, dispersed and irregularly shaped cluster. It is worth noting that a 2° field of view is need; otherwise its resemblance to a cluster is lost. Nevertheless it is worth seeking out.

Notes

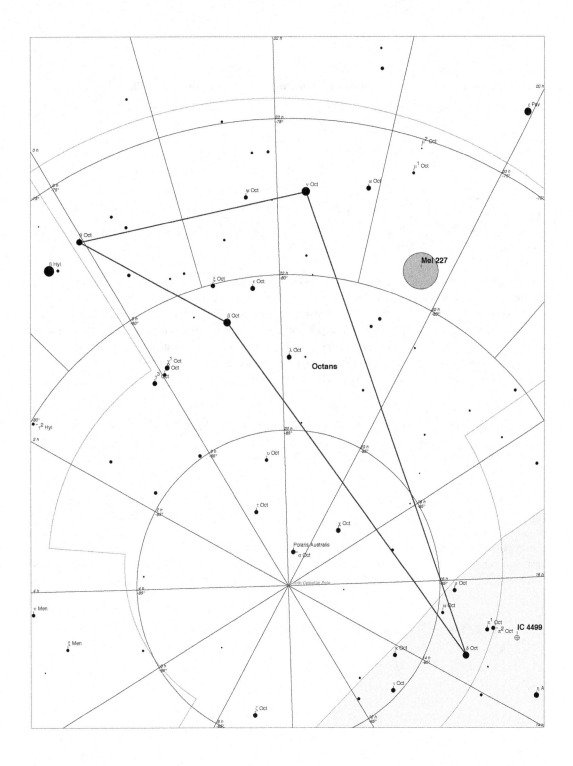

Ophiuchus

Fast Facts

Abbreviation: Oph	Genitive: Ophiuchii	Translation: The Serpent Holder
Visible between latitudes 80° and −80°		Culmination: June

Clusters

NGC 6171	Herschel VI-40	16ʰ 32.5ᵐ	−13° 03′	GC
8.85 m	⊕ 13′		X	Moderate

Also known as Messier 107, this is often missing from amateurs' observing schedules owing to its faintness. It is nevertheless a pleasant cluster with a mottled disc and brighter core. Not visible with the naked eye, it nevertheless presents a pleasing aspect when medium to high magnification is used. What makes this inconspicuous globular important, however, is that it is one of the very few that seem to be affected by the presence of interstellar dust. Deep imaging has revealed several obscured areas within the cluster, possibly due to dust grains lying between us. This isn't such a surprise, as the globular is located over the hub of the galaxy in Scorpius.

NGC 6218	Messier 12	16ʰ 47.2ᵐ	−01° 57′	GC
7.68 m	⊕ 16.0′		IX	Moderate

A small cluster that will be a challenge to naked-eye observers, but in telescopes of aperture 20 cm and more, this cluster is very impressive, with many stars being resolved against the fainter background of unresolved members. It also contains many faint-colored stars that show up well with telescopes of aperture 10 cm and greater. It is nearly the twin of Messier 10, which is within 3° southeast. A recent study has shown that it has a small number of low mass stars, and it has been suggested that many may have been stripped from it by the gravitational influence of the Milky Way.

NGC 6235	Herschel II-584	16ʰ 53.4ᵐ	−22° 11′	GC
8.9 m	⊕ 5.0′		X	Moderate

A low brightness globular cluster that could pose a problem to see with a light-polluted sky. With a small aperture it will be an unresolved diffuse and hazy glow that may appear to have a slight mottled aspect when averted vision is used. At medium aperture it will remain a small hazy spot, and it is only with a large aperture will any resolution of its members be seen, and even then only at its edge with only a handful of stars.

NGC 6254	Messier 10	16^h 57.1^m	–04° 06′	GC
6.4 m	⊕ 20′		VII	Easy

Similar to Messier 12, it is however slightly brighter and more concentrated. Can be seen with the naked eye on dark nights. It lies close to the orange star 30 Ophiuchi (spectral type K4, magnitude 5), and so if you locate this star, then by using averted vision Messier 10 should be easily seen. With apertures of 20 cm and more, the stars are easily resolved right to the cluster's center. Under medium aperture and magnification, several colored components have been reported: a pale blue tinted outer region surrounding a very faint pink area, with a yellow star at the cluster's center.

NGC 6266	Messier 62	17^h 01.2^m	–30° 07′	GC
6.5 m	⊕ 14′		IV	Easy

A very nice cluster, visible in binoculars as a small hazy patch of light set in a wonderful star field. Owing to its irregular shape, it bears a cometary appearance, which is apparent even in small telescopes. Has a very interesting structure where concentric rings of stars have been reported by several observers, along with a colored sheen to its center, described as both pale red and yellow!

NGC 6273	Messier 19	17^h 02.6^m	–26° 16′	GC
6.8 m	⊕ 17′		VIII	Easy

A splendid albeit faint cluster when viewed through a telescope, it nevertheless can be glimpsed with binoculars, where its egg shape is very apparent. Although a challenge to resolve, it is nevertheless a colorful object, reported as having both faint orange and faint blue stars, while the overall color of the cluster is a creamy white. Some amateurs also claim that a few faint dark patches mottle the cluster; perhaps this is interstellar dust between the cluster and us.

NGC 6284	Herschel VI-11	17^h 04.5^m	–24° 46′	GC
8.9 m	⊕ 6.2′		IX	Moderate

Another of Ophiuchus' small and faint globular clusters. Once again the use of averted vision will improve its appearance, as will a modest increase in magnification. It will remain unresolved in all except the largest apertures, and then a few stars of magnitude12 and fainter can be seen in its halo.

NGC 6287	Herschel II-195	17^h 05.1^m	−22° 42′	GC
9.2 m	⊕ 4.8′		IV	Moderate

This globular could be a difficult object to locate under an urban sky so a dark sky may be a pre-requisite. With small aperture it will look like a small circular glow with perhaps a hint of a mottled appearance, but this will depend on the sky conditions, and averted vision will help. With a medium aperture it will just appear a somewhat brighter, unresolved spot. Larger apertures will give a definite granular feel to the core, with a few resolved stars in its halo. If you have access to a very dark sky, try to see the dark dust cloud that lies to its north that seems to block out the starlight.

NGC 6293	Herschel VI-12	17^h 10.2^m	−26° 35′	GC
8.3 m	⊕ 8.2′		IV	Moderate

This globular cluster is just north of the famous dark nebula, Barnard 59, and will appear in small apertures as a small and tight ball of light. The core will have a granular texture and a few outlying stars in the halo will be resolved in medium and large apertures.

NGC 6304	Herschel I-147	17^h 14.5^m	−29° 28′	GC
8.3 m	⊕ 8.0′		VI	Moderate

This is a smallish but bright cluster with only a few resolvable stars near its edge. Nevertheless it will be a challenge to locate with binoculars, and in small telescopes you will need a dark sky and medium magnification.

NGC 6316	Herschel I-45	17^h 16.6^m	−28° 08′	GC
8.1 m	⊕ 5.4′		X	Moderate

Easily recognizable as a globular, this is a small and condensed object. The core will remain unresolved at medium aperture and in larger telescopes will just appear mottled. Not a very impressive object.

NGC 6333	Messier 9	17^h 19.2^m	−18° 31′	GC
7.7 m	⊕ 12.0′		VII	Easy

Visible in binoculars, this is a small cluster, with a brighter core. The cluster is one of the nearest to the center of our galaxy, and is in a region conspicuous for its dark nebulae, including Barnard 64; it may be that the entire region is swathed in interstellar dust, which gives rise to the cluster's dim appearance. It lies about 26,000 light years away.

NGC 6342	Herschel I-149	17^h 21.2^m	−19° 35′	GC
9.5 m	⊕ 4.4′		IV	Difficult

If you like a challenge, then this globular cluster is for you as even a hint of light pollution may make it difficult to locate. Small apertures will only show a tiny spot and increasing both magnification and aperture will only enhance the view slightly. The largest apertures will show a granular center.

NGC 6356	Herschel I-48	17^h 23.6^m	−17° 49′	GC
8.2 m	⊕ 10′		II	Easy

In a constellation that has a lot of faint objects, it is nice to see a fairly bright one, and this globular cluster is no exception. With a dark sky this is quite a nice object, and shows up particularly well in small apertures where it has been compared to Messier 9. A slight granulation of the core can be seen with averted vision, and with an increase in aperture this granularity increases but will still remain unresolved.

NGC 6355	Herschel I-46	17^h 24.0^m	−26° 21′	GC
8.6 m	⊕ 4.2′		–	Moderate

We have another easily seen globular cluster, although it is small and will remain unresolved in small telescopes. Once again, like so many of the globular clusters in Ophiuchus it will only begin to show resolvable stars in large apertures.

NGC 6366	Herschel I-48	17^h 27.7^m	−05° 05′	GC
9.5 m	⊕ 8.3′		XI	Moderate

This low surface globular cluster is weakly concentrated and in medium apertures will show granularity on the limit of resolution, whereas increasing the apertures will of course show more fainter stars in the halo.

Trumpler 26	Harvard 15	17^h 28.5^m	−29° 29′	OC
9.5 m	⊕ 17′	40	II 1 m	Easy

Now here's a nice surprise – we have a rare open cluster. It has about one dozen bright stars and many more faint stars. It is a well-scattered cluster and quite a nice object to observe.

NGC 6402	Messier 14	17^h 37.6^m	–3° 14′	GC
7.6 m	⊕ 11′		VIII	Easy

Located in a sparse region of the Milky Way, abundant in dust, we have a fairly bright, but unresolvable in small apertures, globular cluster. It has been reported as having a very faint yellow hue, maybe due to the interstellar dust that lies between it and us. Increasing aperture will reveal areas of resolved stars, and in fact at high magnifications a distinct orange color can be glimpsed at its core.

NGC 6401	Herschel I-44	17^h 38.6^m	–23° 54′	GC
7.4 m	⊕ 4.8′		VIII	Moderate

Set in a wonderful part of the Milky Way is this small and concentrated cluster. In small apertures and magnification it will just look like an enhancement of the star field, and will only become apparent as a globular cluster by increasing magnification. It has a slight concentration at its core, but will remain unresolved in all apertures.

IC 4665		17^h 46.3^m	+05° 43′	OC
4.2 m	⊕ 40′	30	III 2 m	Easy

A naked-eye object under perfect seeing conditions, this large cluster appears as a hazy spot measuring over two full Moon diameters. With binoculars, nearly 30 blue-white 6th magnitude stars can be seen. Its position in a sparse area of the sky emphasizes the cluster, even though it is not a particularly dense collection of stars.

Collinder 350	–	17^h 48.1^m	+01° 18′	OC
6.1 m	⊕ 45′	20	IV 1 p	Easy

Here we have one of those large clusters that can be seen in a finderscope, but when observed through a telescope can lose its cluster appearance. It is a naked-eye object under clear skies, and in binoculars will consist of about 20 or more stars. In medium and larger apertures some two dozen stars will be seen with many star chains. It can be seen simultaneously with the much fainter star clusters IC 4756 and IC 4665.

NGC 6633	Herschel 72	18^h 27.7^m	+06° 34′	OC
4.6 m	⊕ 27′	25	III 2 m	Easy

Bordering on naked-eye visibility, although reports are hard to come by, this bright, large, but loose cluster is perfect for binoculars and small telescopes. The stars are a lovely bluish-white set against the faint glow of the unresolved members. At the northern periphery of the cluster is a small but nice triple star system.

Notes

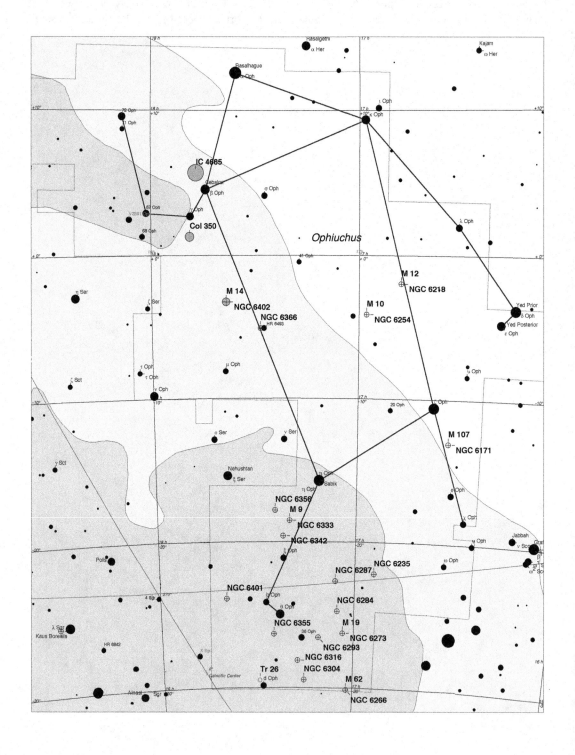

Notes (cont.)

Orion

Fast Facts

Abbreviation: Ori	**Genitive: Orionis**	**Translation: The Great Hunter**
Visible between latitudes 85° and −75°		**Culmination: December**

Clusters

NGC 1662	**Collinder 55**	**04ʰ 48.5ᵐ**	**+10° 56′**	**OC**
6.4 m	⊕ 20′	35	II 3 m	Easy

This open cluster can be seen in binoculars as a bright and large cluster, consisting of just a few stars of around 8th and 9th magnitude. Increasing aperture will reveal 20–30 fainter stars, several of which appear as nice doubles.

Collinder 69	**Lambda Orionis Association**	**05ʰ 35.1ᵐ**	**+09° 56′**	**OC**
2.8 m	⊕ 65′	20	II 3 p n	Easy

This open cluster is a perfect object for binoculars. The cluster surrounds the 3rd magnitude stars λ Orionis, and includes ϕ^{-1} and ϕ^{-2} Orionis, both of 4th magnitude. Encircling the cluster is the very faint emission nebula Sharpless 2–264, but it is only visible when using averted vision and an OIII filter with extremely dark skies.

NGC 1980	**Collinder 37**	**05ʰ 35.4ᵐ**	**−05° 55′**	**OC**
2.5 m	⊕ 14′	30	III 2 p n	Easy

Lying on the edge of the Orion Nebula, Messier 42 is the open cluster NGC 1980 consisting of about 30 stars. It is easy to find because at its center is the wonderful triple star τ Orionis (44 Orionis). This triple has a bluish-white primary and white secondary along with a third red star. Research suggest that the two brighter members were originally associated with different stars, and due to a close encounter these runaway stars were flung outward. Possible candidates for these two runaway stars are AE Aurigae and μ Columbae. The cluster is itself surrounded by nebulosity. But what is truly spectacular is that both the cluster, the triple star, Messier 42 and Messier 43 can all be seen in a one-degree field. Amazing!

NGC 1981	**Collinder 73**	**05ʰ 35.2ᵐ**	**−04° 26′**	**OC**
4.6 m	⊕ 25′	20	III 2 p n	Easy

Now for a bright and coarse open cluster that lies about 1° north of Messier 42. Around eight or nine stars can be seen in binoculars, while the remaining stars are a hazy background glow. In moderate telescopes, the most striking feature is two parallel rows of stars.

NGC 2112	Collinder 76	05^h 53.8^m	−00° 24′	OC
8.4 m	⊕ 18′	50	II 3 m n	Moderate

This is a faint open cluster that needs a moderate aperture, say, 20 cm, to be seen, and only then it will contain about a dozen 12th and 13th magnitude star, along with a 10th magnitude star. Larger apertures will show many more faint members set against a haze of unresolved stars. One added bonus is that the nebula Barnard's Loop[4] passes west of the cluster heading south.

NGC 2141	Collinder 79	06^h 02.9^m	+10° 27′	OC
9.4 m	⊕ 10′	100	I 2 r	Moderate

This is another faint open cluster that needs a moderate aperture in order to be seen. It will appear as just a small group of 13th magnitude stars set in an unresolved speckled haze. Larger apertures reveal a rich cluster and resolve several more stars.

NGC 2169	Collinder 38	06^h 08.6^m	+13° 58′	OC
5.9 m	⊕ 7′	30	I 3 p n	Easy

This is a small but bright open cluster. Some observers find it hard to believe that this scattering of stars has been classified as a cluster. Easily visible in binoculars, the stars appear to range in magnitude from about 8 to 10. Also, binoculars will show the four brightest members to be surrounded by faint nebulosity – sometimes! Near the center of the cluster is the variable star GSC 00742–02169, magnitude 10.8, one of the peculiar Ap stars. The variability is believed to be due to the fact that the axis of rotation and axis of the magnetic field poles are misaligned.

NGC 2194	Collinder 87	06^h 13.8^m	+12° 48′	OC
8.5 m	⊕ 10′	50	II 2 r	Moderate

---

[4] I believe that this nebula presents one of the greatest challenges for a naked-eye observer.

There are several unconfirmed reports that this open cluster has been glimpsed with binoculars. It really needs moderate to large apertures for it to be appreciated. In the former it appears rich and circular with perhaps the hint of an X-shaped asterism at its center. In the latter, many more stars are resolved set against an unresolved glow. Observers suggest that the cluster is best seen with averted vision.

Collinder 70	Orion's Belt	–	–	OC
–m	⊕ 150′			Easy

You may think it odd that the stars in the belt of Orion are included here, but in fact the area of the sky covered by the three 2nd magnitude stars actually contains a impressive sprinkling of stars, ranging from 5th to 8th magnitude and even fainter, with a plethora of barely resolved stars. The whole group has been cataloged as the open cluster Collinder 70. Naturally the three belt stars can be seen with the naked eye, but in binoculars the whole area presents a superb spectacle. This really is a wonderful part of the sky and, sadly, often overlooked. Try it for yourself on cold, crisp, winters evenings. You'll be pleasantly surprised!

The Orion Association

This association includes most of the stars in the constellation down to 3.5 magnitudes, except for γ Orionis and π³ Orionis. Also included are several 4th, 5th and 6th magnitude stars. The wonderful nebula M42 is also part of this spectacular association. Several other nebulae (including dark, reflection and emission nebulae) are all located within a vast giant molecular cloud, which is the birthplace of all the O- and B-type supergiant, giant and main-sequence stars in Orion. The association is believed to be 800 light years across and 1,000 light years deep. By looking at this association, you are in fact looking deep into our own spiral arm that, incidentally, is called the Cygnus-Carina arm.

Notes

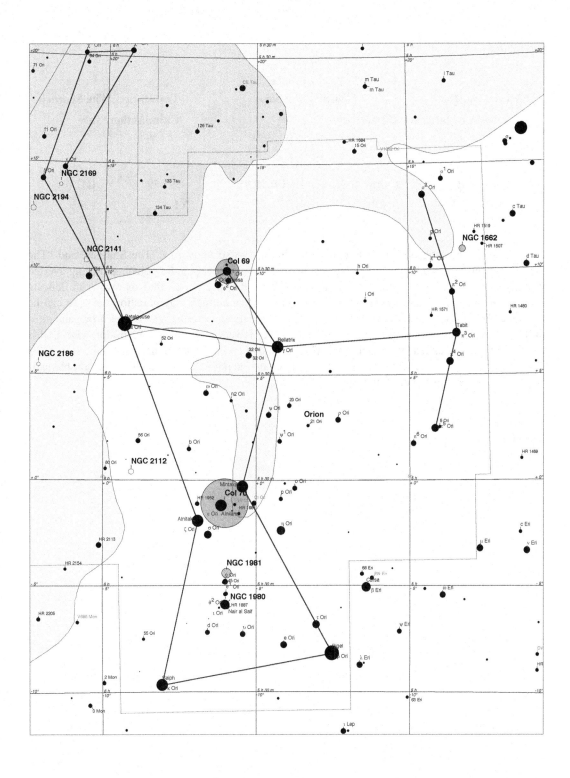

Pavo

Fast Facts

Abbreviation: Pav	Genitive: Pavonis	Translation: The Peacock
Visible between latitudes 15° and –90°		Culmination: July

NGC 6752	Caldwell 93	19^h 10.9^m	–59° 59′	GC
5.4 m	⊕ 20.4′		XI	Easy

The third brightest globular cluster in the entire sky. Also known as "The Starfish" and "The Windmill," this is a lovely naked-eye cluster. In binoculars it is a splendid object, and using averted vision greatly enhances the view. With larger apertures it really comes into its own with delicate streams and chains of stars radiating outward. With a high enough magnification the core can be resolved. This is a very impressive object, so it is a shame it can never be seen from Europe and most of the United States. Research indicates that the cluster is very old at nearly 12 billion years and is undergoing what is called "core collapse." This occurs when the gravitational pull of the stars in its core is so immense; it causes the stars to attempt to converge in on themselves. The once problematical blue stragglers have been found in the cluster.

Notes

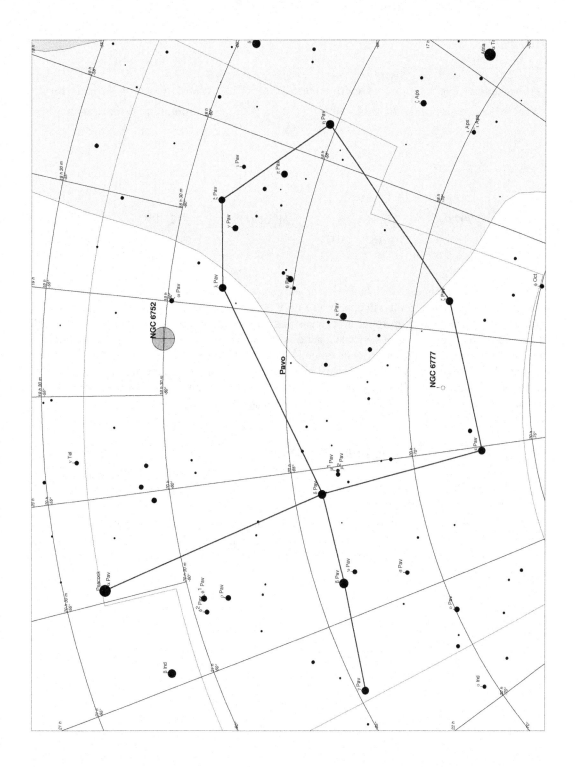

Pegasus

Fast Facts

Abbreviation: Peg	Genitive: Pegasi	Translation: The Winged Horse
Visible between latitudes 90° and −60°		Culmination: August/September

Clusters

NGC 7078	Messier 15	21ʰ 30.0ᵐ	+12° 10′	GC
6.2 m	⊕ 18′		IV	Easy

Messier 15 is an impressive globular cluster and can be glimpsed with the naked eye under perfect conditions. Using binoculars it will appear as a hazy object with no resolvable stars. However, under medium magnification and aperture, it becomes much more impressive and will show considerable detail such as dark lanes, arcs of stars and a noticeable asymmetry. It is one of only four globular clusters that have a planetary nebula located within it – Pease-1. To see this nebula an aperture of 30 cm at least will be needed, along with a detailed star map of the field, the judicious use of an appropriate filter, and a lot of patience. The cluster has undergone core collapse, resulting in an extremely high number of stars at its center that may be home to a black hole. The cluster is also an X-ray source.

NGC 7772	Lund 1049	23ʰ 52.8ᵐ	+16° 15′	OC/Ast
5.0 m	⊕ 5′		III 1 p	Moderate

This is a mediocre open cluster whose inclusion here is warranted only by the lack of anything else (regarding clusters) in Pegasus. Apertures of 20 cm will only show six or seven 13th and 14th magnitude stars in an area of about 5′. Several observers (and a few catalogs) now refer to this as an asterism and I concur with this.

Notes

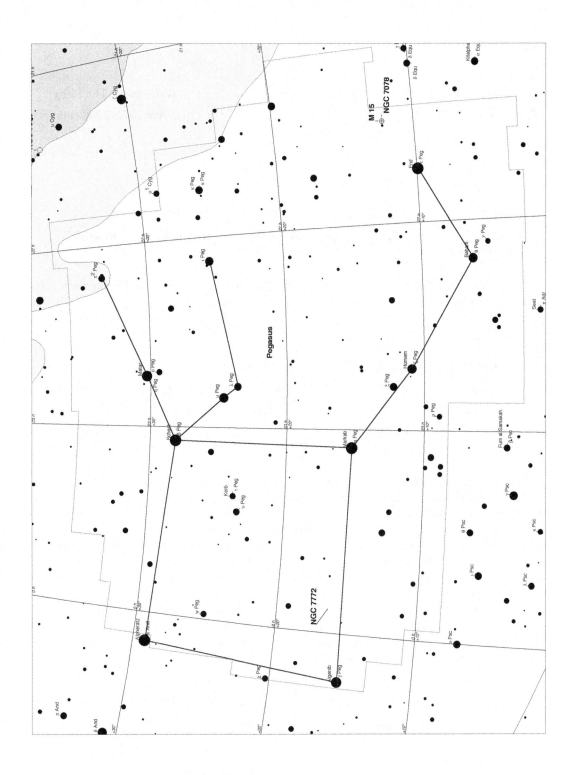

Perseus

Fast Facts

Abbreviation: Per	**Genitive: Persei**	**Translation: The Hero**
Visible between latitudes 90° and −35°		**Culmination: November**

Clusters

NGC 744	Collinder 22	01ʰ 58.5ᵐ	+55° 28′	OC
7.9 m	⊕ 11′	20	III 1 p	Moderate

A spread out and scattered cluster, best seen in moderate to larger apertures. Its brightest star, SAO 22809, is magnitude 7.8 at the clusters north-northeast boundary. The remaining stars are all 10th magnitude and fainter. There isn't much more one can say about this cluster.

NGC 869	Collinder 24	02ʰ 19.0ᵐ	+57° 09′	OC
5.3 m	⊕ 29′	200	I 3 r	Easy
NGC 884	–	02ʰ 22.4ᵐ	+57° 07′	OC
6.1 m	⊕ 29′	115	II 2 p	Easy

Glorious! The famous Double Cluster in Perseus should be on every amateur's observing schedule and is a highlight of the northern hemisphere winter sky. Strangely, it was never cataloged by Messier, even though it is visible to the naked eye, but it is best seen using a low-power, wide-field optical system. But whatever system is used, the views are marvelous. NGC 869 has around 200 members, while NGC 884 has about 150. Both are composed of A-type and B-type supergiant stars with many nice red giant stars. However, the systems are dissimilar; NGC 869 is 5.6 million years old (at a distance of 7,200 light years), whereas NGC 884 is younger at 3.2 million (at a distance of 7,500 light years). But be advised that in astrophysics, especially with distance and age determination, there are very large errors! Also, it was found that nearly half the stars are variables of the type Be, indicating that they are young stars with possible circumstellar discs of dust. Both are part of the Perseus OB1 Association from which the Perseus spiral arm of the galaxy has been named. Don't rush these clusters; spend a long time observing both of them and the background star fields. Also known as Caldwell 14.

NGC 957	Collinder 28	02ʰ 33.3ᵐ	+57° 34′	OC
7.6 m	⊕ 11′	60	III 2 m	Easy

A moderately rich cluster, though somewhat faint and small, nevertheless it can be seen in small telescopes, and actually in all apertures is a nice object, full of double stars, including a particularly fine one at its southern, southwestern corner, consisting of blue and white components.

Trumpler 2	Collinder 29	02^h 36.7^m	+55° 56′	OC
5.9 m	⊕ 20′	20	III 2 p	Easy

The cluster can be glimpsed in a finderscope and is a large open cluster that lays on the line between eta (ε) Persei and the famous Double Cluster. It has several arcs and chains of stars and looks even better in a large aperture telescope. As a bonus, the star eta (ε) Persei is a double with a lovely color contrast of a blue and gold, at magnitudes 8 and 9, respectively.

NGC 1039	Collinder 31	02^h 42.0^m	+42° 47′	OC
5.2 m	⊕ 35′	60	II 3 m	Easy

A nice cluster easily found, about the same size as the full Moon. Messier 34, as it is also known, can be glimpsed with the naked eye and is best seen with medium-sized binoculars, as a telescope will spread out the cluster and so lessen its impact. At the center of the cluster is the double star H1123, both members being 8th magnitude and of type A0. The pure-white stars are very concentrated toward the cluster's center, while the fainter members disperse toward its periphery. Thought to be about 200 million years old, lying at a distance of 1,500 light years.

NGC 1245	Herschel VI-25	03^h 14.7^m	+47° 14′	OC
8.4 m	⊕ 12.0′	200	III 1 r	Moderate

With a small telescope, this will look small and faint with set against a background haze of unresolved member stars. To fully appreciate the cluster, larger apertures are better, and in fact, it becomes an impressive sight with several star streams becoming apparent.

Melotte 20	Collinder 39/40	03^h 22.1^m	+49°	OC
1.2 m	⊕ 185′	50	III 3 m	Easy

If you have a clear, cold winter's night, then this object will be visible as a brightening of the Milky Way, and use of binoculars will show a wonderful star field, lying between alpha (α) and delta (δ) Persei, believed to be among its most outlying members as well as 29 and 34 Persei. This is the Alpha Persei Moving Group. Believed to have over 100 stars, this cluster is obviously best suited to binoculars or rich field telescopes. In fact, even under polluted skies, and a partial Moon, it should still be visible in optical equipment. Its distance has been estimated to lie between 557–650 light years and is around 50–70 million years old. This is a nice, easy to find object, so make sure you seek it out. The inner region of the stream is measured to be over 33 light years in length, the distance between 29 to ψ Persei.

NGC 1342	Herschel VIII-88	03ʰ 31.7ᵐ	+37° 22′	OC
6.7 m	⊕ 17′	40	III 2 m	Easy

Visible in binoculars under a dark sky, of course, this is a bright but scattered cluster. The stars range in magnitudes from 9th to 14th, scattered in chains and loops. Look out for two 8th magnitude stars off the northeastern side that are most likely field stars, along with several prominent star lanes using low magnification, including a rather long stream oriented west to east.

IC 348	Collinder 41	03ʰ 44.6ᵐ	+32° 17′	OC/Neb
7.3 m	⊕ 7′	20	IV 2 p n	Moderate

We now come to a very nice object consisting of a star cluster, IC 348, the reflection nebula, IC 348, and the dark nebulae, Barnard 3 and 4. It would be more accurate to say the IC 468 is not three separate objects, but rather a star-forming region, with three components. Large apertures will of course be needed here to see all parts, as well as judicious use of a filter in order to fully appreciate the nebula. The cluster itself is rather faint and small, and overshadowed by the nebulae. Coincidentally, the dark nebulae are believed to be the remains of the material that formed the Zeta Persei Association, which we will discuss later in this section.

NGC 1444	Herschel VIII-80	03ʰ 49.3ᵐ	+52° 39′	OC
6.6 m	⊕ 4.0′	10	IV 1 p	Easy

A 7th magnitude star lies in this cluster, so it is relatively easy to locate, even in binoculars. Small apertures will show a hazy, unresolved blob, but use of averted vision will resolve areas of it into stars. Increasing aperture will just show more of this small cluster.

NGC 1496	Collinder 44	04ʰ 04.4ᵐ	+52° 38′	OC
9.6 m	⊕ 6′	15	III 2 p	Moderate

This is a small and indifferent cluster, even in moderate to large aperture telescopes. It does however have two nice features. There is a very obvious asterism in the shape of a semicircle, and the cluster is divided into two distinct groups; one is to the west-northwest and the other is to the east-southeast.

NGC 1513	Herschel VII-60	04ʰ 09.9ᵐ	+49° 31′	OC
8.4 m	⊕ 9.0′	50	II 1 m	Moderate

This a is a well spread out and faint object that will prove a challenge to see from an urban environment; thus a large aperture will be needed. However, if you are able to observe this under dark skies, it will look good in a small telescope. But note that when using a small telescope, along with a low magnification, it will just be a faint hazy glow, so a higher magnification will be needed to resolve some of its brighter members.

NGC 1528	Herschel VII-61	04ʰ 15.3ᵐ	+51° 41′	OC
6.4 m	⊕ 23′	40	II 2 m	Easy

Under a dark sky this would be a good naked eye challenge and is a marvelous sight in all types and sizes of optical equipment. Small binoculars will reveal several cluster members, and increasing aperture will reveal lots of star arcs and chains. A delightful cluster.

NGC 1545	Herschel VIII-85	04ʰ 20.9ᵐ	+50° 15′	OC
6.2 m	⊕ 12′	60	IV 2 p	Easy

There are a few reports that this is a naked eye object, but not many have been able to glimpse it, and in binoculars only a few stars can be seen against an unresolved haze. Thus small telescopes (and larger) are needed to appreciate it fully. An increase of magnification will naturally reveal more stars, 9th magnitude and fainter, and using averted vision will bring many of the fainter stars into sight, but it does lie in a rich part of the Milky Way, so the edge of the cluster melds into the background star field. It does however contain a few nicely colored double stars.

NGC 1582	Collinder 51	04ʰ 32.0ᵐ	+43° 51′	OC
7.0 m	⊕ 37′	25	IV 2 p	Easy

This is a large and scattered cluster of stars with magnitudes ranging from 9th to 13th. What makes it worth observing is that it has many arcs and chains of stars, along with a very impressive stream of fairly bright stars almost 20′ long at its southwest.

Berkeley 68	–	04ʰ 44.5ᵐ	+42° 04′	OC
9.8 m	⊕ 10′	60	IV 2 p	Difficult

This is a cluster that is not often seen in guidebooks, or even mentioned anywhere else for a couple of reasons. It is from the Berkeley Open Clusters program, devised by astronomers at the University of Berkeley in 1958. The cluster is barely resolved, even in large apertures with stars of 13.5 magnitude and fainter, so you can appreciate the problem of finding this one. To add to its woes, it is also in a relatively rich part of the Milky Way. Nevertheless, seek this one out, as it is not often visited.

The Zeta Persei Association

Once again here is one of those enormous but very interesting stellar associations you can add to your observing list. Also known as Per OB2, this association includes z and x Persei, as well as 40, 42 and o Persei. The California Nebula, NGC 1499, is also within this association. It lies at a distance of about 1,300 light years.

Notes

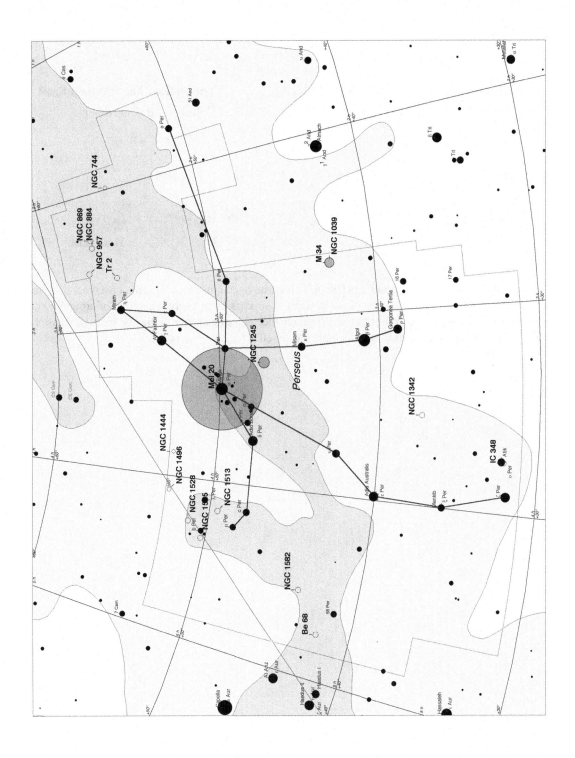

Pictor

Fast Facts

Abbreviation: Pic	**Genitive: Pictoris**	**Translation: The Painter's Easel**
Visible between latitudes 25° and –90°		**Culmination: December**

Cluster

NGC 2132	–	05ʰ 55.2ᵐ	–59° 55′	OC/Ast
–m	⊕ 45′	15	–	Moderate

The only cluster in Pictor poses a problem. Is it a true open cluster or, as some observers believe, just an asterism? In a large aperture of, say, 30 cm, it appears as an enormous group of around 50 or so stars, but even then it has only a minimal resemblance to a true cluster. It may be more of an asterism, but you should have a look and decide for yourself.

Notes

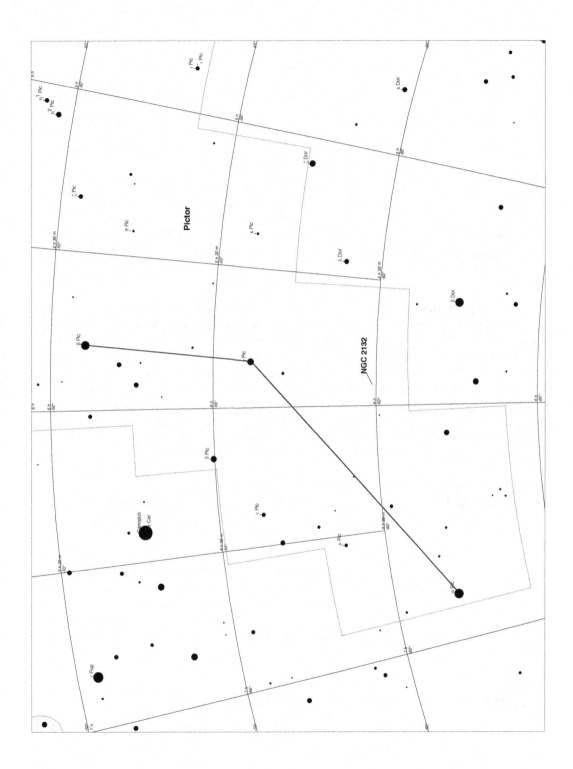

Puppis

Fast Facts

Abbreviation: Pup	**Genitive: Puppis**	**Translation: The Stern**
Visible between latitudes 40° and –90°		**Culmination: January**

Clusters

NGC 2298	–	06ʰ 49.0ᵐ	–36° 00′	GC
9.3 m	⊕ 6.8′		VI	Easy

Located in a very rich part of the Milky Way, this globular cluster is visible in moderate apertures as a smooth small patch of light with no resolution of stars. As aperture is increased a brighter core emerges with a faint halo, along with several stars being resolved in the core, and half a dozen more in the outer halo. What makes this fairly ordinary looking cluster interesting, however, is the idea that it may in fact not be a member of our home galaxy, but is associated with the Canis Major dwarf galaxy that is currently being cannibalized by the Milky Way. The tidal disruption has resulted in a long filament of stars trailing behind it as it orbits the Milky Way, forming a complex ring-like structure referred to as the Monoceros Ring, which wraps around our galaxy three times. Other clusters include NGC 1851, NGC 1904, NGC 2298 and NGC 2808, all of which are likely to be remnants of the galaxy's globular cluster system before its accretion into the Milky Way.

Collinder 135	–	07ʰ 17.0ᵐ	–36° 50′	OC
2.1 m	⊕ 50.0′		IV 2 p	Easy

Also known as the Pi Puppis cluster, this cluster is actually visible to the naked eye. In binoculars and small telescopes what becomes immediately apparent are the four bright stars that form an equilateral triangle. The star Pi (π) Puppis is a nice double, a pale yellow primary and much fainter bluish secondary. Several other fainter stars are scattered irregularly. This is a nice, often overlooked cluster, and to clear up the many arguments that abound, Collinder 135 is a true cluster, and Pi Puppis is a member of that cluster.

Melotte 66	Collinder 147	07ʰ 26.4ᵐ	–47° 40′	OC
7.8 m	⊕ 10′	40	II 1 r	Moderate

This is a fairly rich, bright, large, and scattered cluster of about 25 stars of 10th magnitude and fainter. A medium or large aperture will be needed to appreciate the cluster. In such telescopes, one will see a background of unresolved stars along with several curved arcs of stars.

NGC 2414	Collinder 150	07^h 33.2^m	−15° 27′	OC
7.9 m	⊕ 4′	30	I 3 m	Moderate

Although this open cluster is faint, it easily stands out from the background haze. It has several 12th magnitude and fainter stars irregularly scattered, along with some nice chains of stars. In a small aperture it will appear as a faint nebulous patch, but using averted vision will resolve a few stars.

NGC 2421	Herschel VII-67	07^h 36.2^m	−20° 37′	OC
8.3 m	⊕ 8.0′	70	I 2 m	Moderate

An open cluster that needs at least a small aperture telescope in order to be observed, this is a fairly small and faint object, albeit rich. With such a telescope it will appear as small haze ball, with perhaps a few barely perceptible outlying stars. Increasing both aperture and magnification will reveal a somewhat richer cluster of about three dozen 11th–13th magnitude stars, grouped in a distinct triangular shape.

NGC 2422	Herschel VIII-38	07^h 36.6^m	−14° 29′	OC
4.4 m	⊕ 25′	50	I 3 m	Easy

Also known as Messier 47, this is a large and wonderful cluster, easily visible to the naked eye, and in binoculars presents a terrific sight with many stars being resolved. This is a big cluster, nearly the same angular size of the Moon, and thus a rich field telescope will give the best views. Increasing aperture will show you star chains, knots and apparently starless voids, along with several bright blue stars. It is a lovely object to observe that will repay repeat visits. Incidentally, there is no cluster in the position given by Messier, which he expressed in terms of its right ascension and declination with respect to the star 2 Puppis. But if the signs of Messier's coordinate differences are corrected, the position matches that of NGC 2422.

Melotte 71	Collinder 155	07^h 37.5^m	−12° 04′	OC
7.1 m	⊕ 9′	60	II 2 r	Easy

This is an open cluster that some observers report looks just like a globular cluster, and in fact it does! It can be seen in large binoculars and can be glimpsed in a finderscope. It consists of about four dozen 11th magnitude stars and fainter that can be resolved, set against a background haze. In addition, some observers report seeing a lot of red stars in the cluster that are probably evolved red giant stars.

NGC 2423	Herschel VII-28	07ʰ 37.6ᵐ	–13° 52′	OC
6.7 m	⊕ 12′	60	II 2 m	Easy

The problem with this cluster is twofold; it is large and scattered and is in a very rich part of the Milky Way, so it tends to meld into the background. Therefore it is best seen with a low magnification so as to encompass the whole cluster in the field of view. Otherwise it will lose its clustering effect. It can be seen in binoculars under a dark sky, but will suffer from the problems noted above.

NGC 2439	Collinder 158	07ʰ 40.8ᵐ	–31° 41′	OC
6.9 m	⊕ 10′	80	II 3 r	Easy

This is a pleasant and rich cluster, well scattered and reasonably bright. It includes the orange star R Puppis, a true member of the cluster. Under good conditions, two double stars may also be seen that seem at right angles to each other.

NGC 2437	Collinder 159	07ʰ 41.8ᵐ	–14° 49′	OC
6.1 m	⊕ 27.0′	30	III 2 m	Easy

Just like its close companion, Messier 47, this cluster, Messier 46, is large and bright. Paradoxically, however, even though this cluster is fainter, it seems to be the better cluster visually, as it is far more concentrated. Visible in binoculars (in the same field of view with Messier 46), it immediately jumps out at you in small and larger apertures, since it is far denser than the background, with the stars distributed equally over the sky. An added bonus is the foreground planetary nebula NGC 2438 that can be seen even in small apertures and will resolve itself into a definite disc at higher magnification and aperture. The American astronomer Phil Harrington has this to say: "NGC 2438 glows softly at 11th magnitude and measures 65″ of arc across. Although their alignment is only by chance, the planetary nebula against the more distant cluster makes M46 one of my favorite treasures."

NGC 2447	Collinder 160	07ʰ 44.5ᵐ	–23° 51′	OC
6.2 m	⊕ 22′	70	I 3 r	Easy

This is a bright, large, and irregular open cluster, also known as Messier 93. It has a definite wedge-shape to it, although it has been reported as being shaped like a starfish. In a large aperture it becomes much more impressive with a few star chains and star voids. Apparently it lies within the Klingon Empire!

NGC 2451	Collinder 161	07^h 45.2^m	−37° 58′	OC
2.8 m	⊕ 45′	40	II 2 m	Easy

Now we come to a very interesting object. Observationally this is a very bright, very large and very irregular cluster that can be seen in a finderscope. It can also be glimpsed with the naked eye under a dark sky. In a small aperture it will appear as a group of around 30 stars from 6th to 11th magnitude. In a larger aperture the cluster becomes very colorful, with blue and yellow stars surrounding the 4th magnitude star c Puppis, which is itself a nice reddish-orange. But the interesting aspect of the cluster is that for a while it was believed not to be a cluster but an asterism, and then at the end of the twentieth century it was suggested that NGC 2451 is in fact two open clusters lying along the same line of sight. This was later confirmed, and the clusters are labeled NGC 2451 A and NGC 2451 B, located at distances of 640 and 1,179 light years, respectively.

NGC 2453	Collinder 162	07^h 47.6^m	−27° 11′	OC
8.3 m	⊕ 5.0′	25	I 3 m	Easy

This open cluster is small and faint, but not prohibitively so; it can be glimpsed under dark skies with a large finderscope and therefore also in binoculars. Most of the stars range in magnitude from 10th to 13th. Other than that, there isn't a lot to say except that the planetary nebula NGC 2452 lies some 6.0′ to its south.

NGC 2477	Collinder 165	07^h 52.2^m	−38° 32′	OC
5.8 m	⊕ 27′	150	I 2 r	Easy

Caldwell 71 is a magnificent cluster that would be far more famous if it were at a higher declination, but alas, it is so far south that most European and American observers cannot see it. It is a naked-eye object if you are lucky enough to see it high in the sky, but if it is too close to the horizon it will be a challenge. It is also quite large, nearly as big as the Moon, and is wonderful in binoculars. It is a terrific sight in small to medium apertures, with a plethora of star chains and loops. In fact, it can resemble a globular cluster such is its richness. It deserves its claim of being one of the top ten open clusters in the entire sky. This is one of those rare objects where any written description doesn't do it justice. Shame more people don't have an opportunity to see it.

NGC 2467	Collinder 164	07^h 52.5^m	−26° 25′	OC/Neb
7.1 p	⊕ 15.0′	40	I 3 m n	Easy

This is an open cluster associated with a nebula, and in fact not many people know of or observe this delightful little object. It is a large and rich cluster that can be glimpsed in a finderscope, and under dark skies it can be seen to be embedded within the nebula. It is believed that the stars are relatively young, having been born out of the material of the nebula, and both are jointly called the Puppis OB2 Association. Incidentally, there are two fainter open clusters to the north of the nebula, Haffner 18 and Haffner 19, the latter of which is now thought to be part of the association. Do try to observe this little object, as it is worth it.

NGC 2479	Herschel VII-58	07ʰ 55.1ᵐ	−17° 43′	OC
9.4 m	⊕ 13.8′	50	III 1 m	Moderate

In a constellation famed for spectacular clusters we now have a rather faint and scattered object that could be difficult to see in binoculars. There are a lot of faint stars at about 12th magnitude set against a background haze of unresolved members. This isn't a particularly impressive cluster. Think of it more as a challenge for binoculars.

NGC 2482	Herschel VII-10	07ʰ 55.2ᵐ	−24° 15′	OC
7.3 m	⊕ 12.0′	40	IV 1 m	Easy

Visible in binoculars under a dark sky, this is a fairly rich but faint cluster that is best suited to small and medium apertures. The stars range from 10th to 12th magnitude. Larger apertures will show a few star chains and loops.

NGC 2489	Herschel VII-23	07ʰ 56.2ᵐ	−30° 04′	OC
7.9 m	⊕ 8.0′	50	I 2 m	Easy

A small but rather nice open cluster awaits the observer here, with a fairly large and bright appearance. It can be glimpsed in a finderscope, with averted vision, so naturally it is visible in binoculars. Actually it is a rather good object in all sizes of aperture, with stars of differing magnitudes arrayed in nice arcs and chains.

NGC 2509	Herschel VIII-1	08ʰ 00.8ᵐ	−19° 03′	OC
93 m	⊕ 12.0′	70	I 1 r	Easy

Another of Puppis' faint clusters, this is fairly faint and somewhat patchy but can be glimpsed in binoculars, although is better, naturally, in telescopes. Small apertures will show a hazy aspect, with a minute central condensation, whereas an increase of aperture will resolve some stars with magnitudes of 10th–13th, with the majority of them giving the impression of either a parallelogram or a ring, depending on how you interpret it.

Collinder 173	–	08ʰ 02.8ᵐ	–46° 23′	Assoc
0.6 m	⊕ 370′	–	II 3 m	Easy

Now for something that will perhaps come as a surprise to most observers. This is an enormous group of stars, covering nearly 5° occupying the entire southern section of the constellation and is identifiable to the naked eye, thanks to the brightness of its stars, which reach the 4th and 5th magnitude. Through binoculars, however, it appears as a vague thickening of the stars, slightly richer than the rest of the sky, and thus it is not possible to see it completely given its large angular size. In actuality, it is an association called the Vela OB2 OB Association located in the nearby Orion arm of the Milky Way.

NGC 2527	Herschel VIII-30	08ʰ 04.9ᵐ	–28° 08′	OC
6.5 m	⊕ 22′	45	III 1 p	Easy

Another binocular object under dark skies, this cluster is quite large and fairly bright. In small telescopes it will appear as a group of around 20 or so stars, but in larger apertures will look more like a scattered and loose collection of three dozen or so 9th–13th magnitude stars set against a rich backdrop. Be aware that in some older observing lists the cluster is given a much smaller diameter of about 10 arcsec.

NGC 2533	Collinder 175	08ʰ 07.1ᵐ	–29° 53′	OC
7.6 m	⊕ 3.5′	20	II 2 r	Easy

A small cluster that will look like a hazy patch with a small aperture telescope; has a 9th magnitude star embedded within it. Even in a moderate or large aperture it is not impressive and somewhat lost in the background star field.

NGC 2539	Herschel VII-11	08ʰ 10.6ᵐ	–12° 49′	OC
6.5 m	⊕ 15.0′	60	II 2 m	Easy

This is a rather bright and large, somewhat oval, open cluster of nearly 100 stars of magnitude 9 and fainter. Under a dark sky, it will just about be a naked-eye object, and in binoculars it will be a hazy spot. The triple star 19 Puppis lies at its southeastern edge, with a yellow primary and two white secondaries. Larger apertures will reveal some star chains and voids. Actually, it is a nice cluster that will repay detailed observation.

NGC 2546	Collinder 178	08^h 12.3^m	−37° 35′	OC
6.3 m	⊕ 40′	40	III 2 m	Easy

Another overlooked cluster, this is a naked-eye object, just, as it lies in a rich star field itself. A very large cluster, bright and scattered, that will fill the field of view of most apertures at low magnification. It also has some nice colored stars: a yellow 6th magnitude star on its southern boundary, and a bluish 7th magnitude star on its northern-northwestern edge. Apparently the blue star is a true cluster member. Well worth seeking out and observing.

NGC 2567	Herschel VII-64	08^h 18.5^m	−30° 39′	OC
7.4 m	⊕ 11.0′	40	II 2 m	Easy

Just like the preceding cluster, this, too, can be a difficult one to observe, for just the same reasons. Visible in binoculars under a dark sky, a telescope is really needed for it to be appreciated, and once located, it has a rather nice string of stars going north to south, easily identified.

NGC 2571	Herschel VI-39	08^h 18.9^m	−29° 45′	OC
7.0 m	⊕ 7.0′	40	II 2 m	Moderate

We now come to a somewhat difficult object to locate in a small aperture, due to it being in a fairly rich star field, and also because it is so large and scattered. Not really visible in binoculars, it appears as an oblong-shaped object with a dark lane bisecting it. Larger apertures will show more star members.

NGC 2579	Collinder 182	08^h 20.9^m	−36° 13′	OC/Neb
7.5 m	⊕ 10.0′	20	IV 2 p	Moderate

Both a cluster and a nebula, this is not a particularly impressive open cluster, overshadowed as it is by the nebula, Gum 11. The cluster is fairly faint, not particularly large, and well scattered, with quite a few 11–13th magnitude members.

NGC 2587		08ʰ 23.4ᵐ	−29° 32′	OC
9.2 m	⊕ 9′	40	III 2 m	Easy

Our final object in Puppis is, alas, not particularly impressive. It is a small, faint and compressed object with stars 12th magnitude and fainter set against an unresolved haze in larger apertures.

Notes

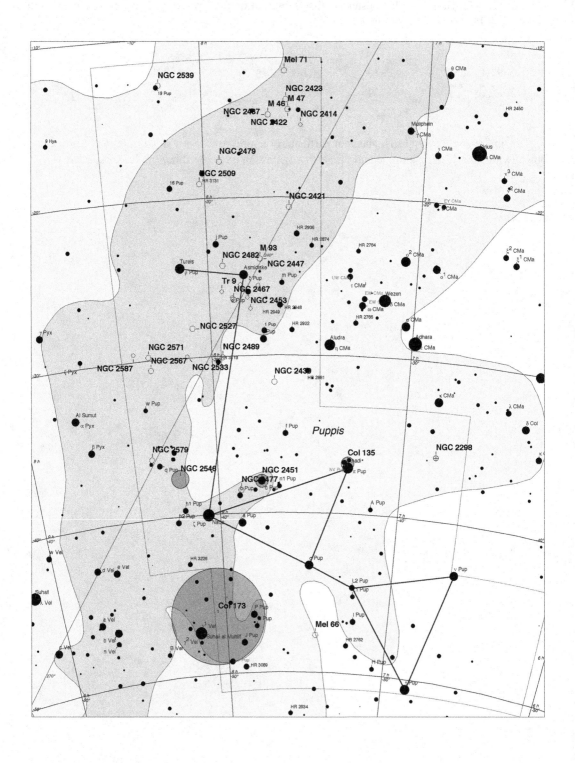

Notes (cont.)

Pyxis

Fast Facts

Abbreviation: Pyx	Genitive: Pyxidis	Translation: The Compass Box
Visible between latitudes 50° and −90°		Culmination: February

Clusters

NGC 2627	Herschel VII-63	08ʰ 37.2ᵐ	−29° 57′	OC
8.4 m	⊕ 11.0′	55	III 2 m	Moderate

With a dark sky this cluster should be visible with binoculars as a faint pale glow. Even in a small aperture, light pollution may make the cluster a problem to observe. However, if glimpsed it will have around 15 or more stars and is a moderately rich object in medium apertures, with around thirty 11th–13th magnitude stars. It does tend to seem more concentrated at its center, set against a background haze of unresolved stars.

NGC 2635	Collinder 190	08ʰ 38.4ᵐ	−34° 46′	OC
11.2 m	⊕ 3′	15	I 3p	Moderate

This is a very small and faint cluster that shows a distinct mottling or granularity to it. Not really visible in a small aperture, it then means medium or large telescopes will be required. Even then it will only show about 10–20 stars, respectively.

NGC 2658	Collinder 195	08ʰ 43.4ᵐ	−32° 39′	OC
9.2 m	⊕ 12′	80	I 2 r	Moderate

To appreciate this cluster it's best to use a medium aperture. This will show about two dozen 12th magnitude stars and an unresolved background haze. With the right conditions several dark lanes may be seen.

NGC 2818	Collinder 206	09ʰ 16.0ᵐ	−36° 38′	OC
8.2 m	⊕ 9′	40	II 2 m n	Moderate

Our final cluster in Pyxis is something of a mystery. Or rather, the object associated with it is. The cluster is fairly faint with about two dozen stars of magnitudes 12–15, therefore is best observed with medium or large aperture telescopes. However, what has caused some discussion is the planetary nebula associated with the cluster, NGC 2818A. For a long time it was assumed that the nebula was embedded within the cluster; however, recent research on the radial velocities of both cluster and nebula suggests it is just a random alignment of objects.

Notes

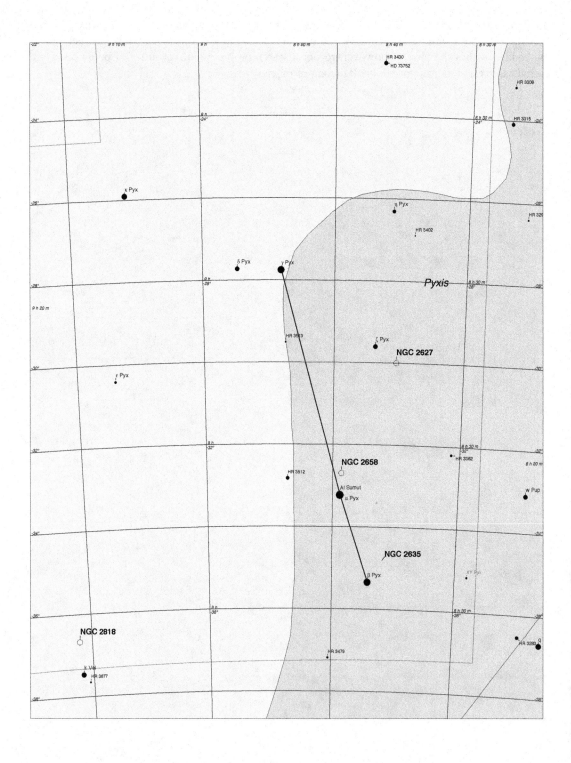

Notes (cont.)

Sagitta

Fast Facts

Abbreviation: Sge	Genitive: Sagittae	Translation: The Arrow
Visible between latitudes 90° and −70°		Culmination: July

Clusters

Palomar 10	–	19ʰ 18.02ᵐ	+18° 34′	GC
13.2 m	⊕ 3.1′		XII	Difficult

Now we have an observing challenge. This globular cluster is often regarded as one of the most challenging Palomar objects to locate and observe. It is faint, small, and even under medium magnification with a large aperture, remains unresolved with just a granular appearance. Nevertheless, have a go and see if you can find it.

Harvard 20	–	19ʰ 53.3ᵐ	+18° 19′	OC
7.7 m	⊕ 8.0′	20	IV 2 p	Moderate

This open cluster is a somewhat difficult binocular object, as the stars are of 12th and 13th magnitude and spread out without any noticeable concentration. It also doesn't help that it is set against the rich star-strewn background of the Milky Way. The two 9th magnitude members are easily spotted, however.

NGC 6838	Messier 71	19ʰ 53.8ᵐ	+18° 47′	GC
6.1 m	⊕ 7.2′		X	Easy

Located in a glittering star field this is a rich and compressed globular cluster that can be glimpsed in a large finder telescope, but will only show a very faint glow in binoculars. But what makes this globular so nice from an observer's point of view is that the central stars can be resolved all the way to the core, which is rare among globular clusters. Increasing aperture will reveal many faint members of the cluster. Up until recently there was some debate as to whether this was a globular or open cluster. The consensus now is that it is a very young globular cluster, only 13,000 light years away, making it one of the closest to us. Also, its orbit around the Milky Way is completely within the disk of the galaxy, and this makes it one of just a few "disk clusters."

NGC 6873	–	20ʰ 08.3ᵐ	+21° 06′	OC
6.4 m	⊕ 12′	22	–	Difficult

Here is another observing challenge for you in this small constellation. This faint and irregular open cluster, if it can even be called that, is cataloged as a "nonexistent" object in the *RNGC*, and that's not surprising, as it will appear to be just a smattering of stars approximately northeast of θ Sagittae. In fact, it can be said with some justification that the best thing about the cluster is the star just mentioned, θ Sagittae, a nice triple star system consisting of a close pair of yellow stars along with an orange companion to its southwest. This star may, or may not, be a member of the cluster.

Notes

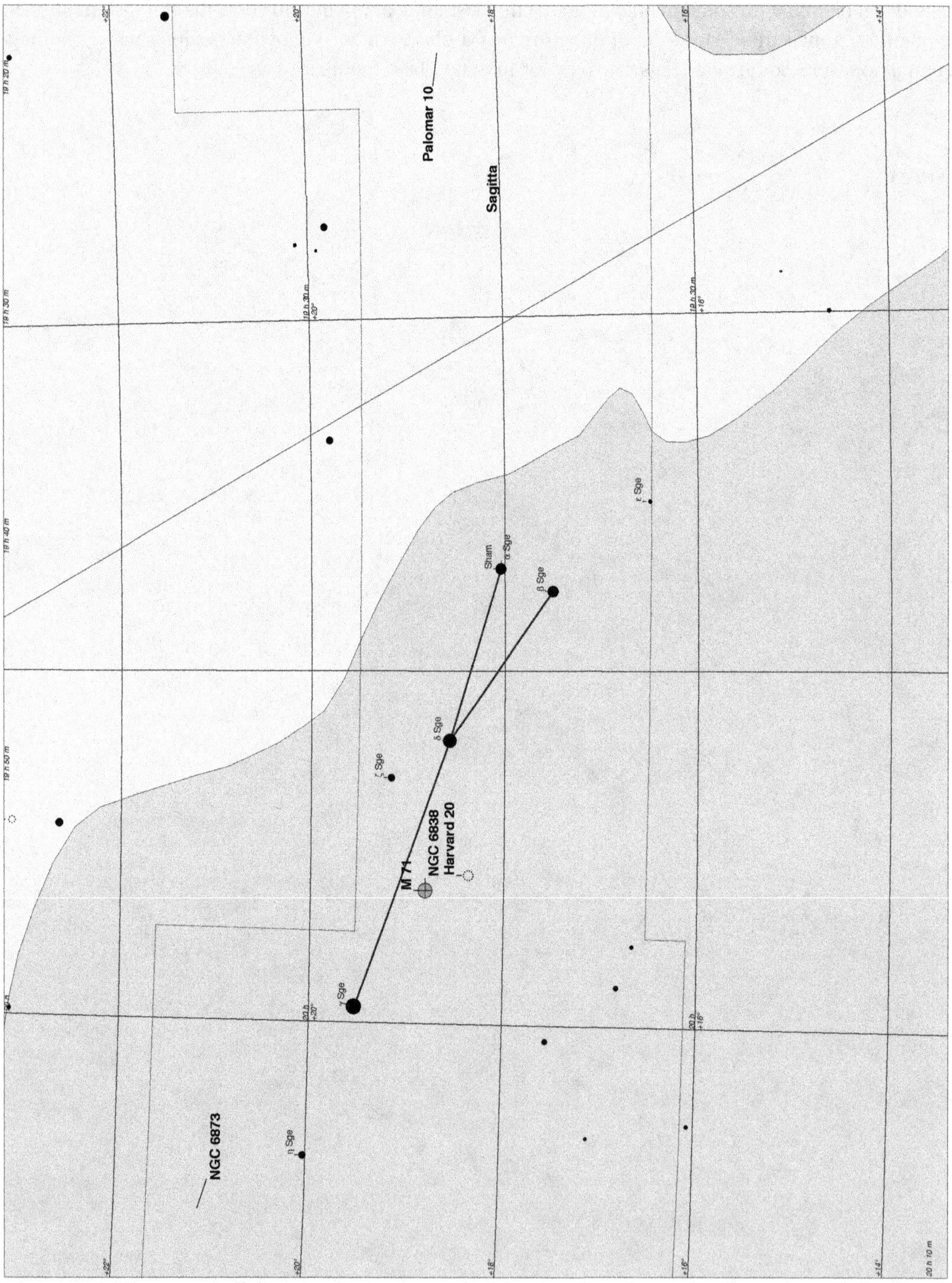
Palomar 10
Sagitta
Sham
α Sge
β Sge
ε Sge
δ Sge
ζ Sge
M 71
NGC 6838
Harvard 20
γ Sge
η Sge
NGC 6873

Notes (cont.)

Sagittarius

Fast Facts

Abbreviation: Sgr	Genitive: Sagittarii	Translation: The Archer
Visible between latitudes 55° and −90°		Culmination: July

Clusters

NGC 6469	–	17ʰ 52.9ᵐ	−22° 21′	OC
8.2 p	⊕ 12′	50	IV 2 m	Difficult

In a constellation that has many wonderful clusters, one will always find a few that are barely clusters at all, and those that some observers have the temerity to call clusters are so faint or indistinct that only the largest apertures will reveal them. Welcome to NGC 6469. This object, even in a large aperture telescope, will just about stand out from the background of the Milky Way. Consider it a test of the acuity of your night vision.

NGC 6494	Collinder 356	17ʰ 56.8ᵐ	−19° 01′	OC
5.5 m	⊕ 27′	100	II 2 r	Easy

Often overlooked because it lies in an area studded with celestial showpieces, Messier 23 is a wonderful cluster that is equally impressive seen in telescopes or binoculars, but the latter will only show a few of the brighter stars shining against a misty glow of fainter stars. The cluster is full of double stars and star chains.

NGC 6520	Herschel VII-7	18ʰ 03.4ᵐ	−27° 53′	OC
7.6 m	⊕ 5′	30	I 2 r n	Difficult

This cluster, although fairly bright, is situated within the Great Sagittarius Star Cloud, and thus makes positive identification difficult. It contains about three dozen faint stars, and locating it is a test of an observer's skill. Nevertheless it is a splendid object, once located. It has a definite elongated shape, and there is a nice red star set among the white stars. Under good conditions, the dark nebula Barnard 86 can be seen to the northwest of the cluster.

NGC 6522	Herschel I-49	18ʰ 03.6ᵐ	−30° 02′	GC
9.9 m	⊕ 9.4′		VI	Moderate

With telescopes of aperture 20 cm this cluster will appear with a bright core but an unresolved halo. A difficult object to locate with binoculars but is rather nice in small telescopes.

NGC 6531	Collinder 363	18ʰ 04.6ᵐ	–22° 30′	OC
5.9 m	⊕ 14′	60	I 3 r	Easy

Also known as Messier 21, this is an outstanding cluster for small telescopes and binoculars. A compact, symmetrical cluster of bright stars with a nice double system of 9th and 10th magnitude located at its center. It lies very close to the Triffid Nebula. In the cluster is the grouping called Webb's Cross, which consists of several stars of 6th and 7th magnitude arranged in a cross. Several amateurs report that some stars within the cluster show definite tints of blue, red and yellow. Can you see them? It is another of those clusters that has not been very well studied.

NGC 6530	–	18ʰ 04.8ᵐ	–24° 20′	OC
4.6 m	⊕ 14′	100	II 2 m n	Easy

Its host nebula – the Lagoon Nebula – often overshadows this open cluster, which would under any other circumstances be easily visible. In a small telescope, the cluster, consisting of about 25 stars, can be found in the northern part of the easternmost area of the nebula. The use of a larger telescope however will show a lovely cluster of glittering points contrasting splendidly with the diffuse nebulosity. Now 40 or more stars can be seen in an area of nearly half a degree. Quite a nice, and often ignored, open cluster.

NGC 6528	Herschel II-200	18ʰ 04.8ᵐ	–30° 03′	GC
9.6 m	⊕ 5.0′		V	Difficult

Even in large telescope of aperture 35 cm, this cluster is unresolved. It will just appear as a faint glow with a slightly brighter center. This would be a good challenge for large-aperture telescopes.

NGC 6546	–	18ʰ 07.2ᵐ	–23° 20′	OC
8.0 m	⊕ 13′	150	II 1 m	Moderate

Lying close to the Lagoon Nebula, this scattered and bright cluster lies in a rich patch of the Milky Way. There are many 11th magnitude stars, and of course many fainter ones. The cluster is evenly scattered, so don't look for any central concentration.

NGC 6544	Herschel II-197	18ʰ 07.3ᵐ	–25° 00′	GC
7.5 m	⊕ 9.2′	10	I 1 p n	Easy

Here we have another cluster that is often overlooked, which is a shame, as it is a nice object. Under dark skies it is visible in binoculars, and in telescopes ranging from small to large, it is a delight. With high magnification and moderate aperture even the halo's outer stars become resolved. Well worth seeking out.

NGC 6553	Herschel IV-12	18ʰ 09.3ᵐ	−25° 54′	GC
8.3 m	⊕ 9.2′		XI	Moderate

Not easily visible in binoculars (although it would prove an observational challenge to locate), it is a fairly even bright cluster, with no perceptible increase in intensity towards the core. If you live in an urban location, with the usual evil of light pollution, it may be difficult to locate in a small telescope.

Collinder 367	–	18ʰ 09.6ᵐ	−23° 59′	OC/?
6.4 m	⊕ 37′	30	III 3 m n	Difficult

This is another irregularly scattered cluster with its members having a range of magnitudes. What makes this cluster interesting is the question – is a cluster at all? As it lies in a rich star field, many observers report that it is all but indistinguishable from the background.

NGC 6568	Herschel VII-30	18ʰ 12.7ᵐ	−21° 35′	OC
8.6 m	⊕ 12′	40	I5 1 m	Moderate

This open cluster really needs a dark sky to be appreciated; otherwise it is a faint and large object, consisting of about 50 stars at around 11th magnitude.

NGC 6569	Herschel II-201	18ʰ 13.6ᵐ	−31° 49′	GC
8.4 m	⊕ 6.4′		VIII	Moderate

This is another cluster that may be difficult to find in an urban setting. In a small telescope with medium magnification, it will appear as a small circular blob, with no discernible variation across it. Only with higher aperture will any gradient of brightness be seen.

Messier 24	–	18ʰ 16.5ᵐ	−18° 50′	–
2.5 m	⊕ 95′ x 35′		–	Easy

Although not a true cluster, it nevertheless must be included in any list. This is the Small Sagittarius Star Cloud, visible to the naked eye on clear nights and nearly four times the angular size of the Moon. It is a superb object for binoculars. The cluster is in fact part of the Norma spiral arm of our galaxy, located about 15,000 light years from us. The faint background glow from innumerable unresolved stars is a backdrop to a breathtaking display of 6th to 10th magnitude stars. It also includes several dark nebulae, which adds to the three-dimensional impression. Many regard the cluster as truly a showpiece of the sky. Spend a long time observing this jewel!

NGC 6595	Collinder 371	18ʰ 17.0ᵐ	−19° 52′	OC/?
7.0 m	⊕ 4′		III 3 m n	Difficult

This is another one of those objects that is classified as both an open cluster as well as a nebula. Many observers report that no cluster exists at the coordinates given, just a double star. What do you see?

NGC 6613	Collinder 376	18ʰ 19.9ᵐ	−17° 08′	OC
6.9 m	⊕ 10′	30	II 3 p n	Easy

A small and unremarkable Messier object, Messier 18, and perhaps the most often ignored, this little cluster, containing many 9th magnitude stars, is still worth observing. Best seen with binoculars or low-power telescopes. A double star is located within the cluster.

NGC 6618	–	18ʰ 20.8ᵐ	−16° 11′	OC
6.0 m	⊕ 11′	40	III 3 m n	Moderate

The cluster itself, consisting of about 35 stars, is in fact located within the glorious Omega Nebula and is well spread out, and so its impact is eclipsed by the magnificence of the nebula. It is these stars that are thought to be responsible for causing the gas to glow, and research indicates that it is one of the youngest clusters known at just 1 million years. A blue hypergiant, HD 1688625, may be a cluster member.

NGC 6624	Herschel I-50	18ʰ 23.7ᵐ	−36° 38′	GC
7.6 m	⊕ 8.8′		VI	Moderate

This is a very small but bright globular cluster that can, under perfect conditions, be glimpsed in binoculars. Oddly, it is brighter than several Messier globular clusters in the constellation, which makes one wonder why Messier left it out of his catalog? Due to its small size, it can be easy to miss this cluster when using a small telescope. In fact, it is wise to use averted vision here to first locate it, and then switch to higher magnification and maybe a larger aperture.

NGC 6652	–	18^h 24.0^m	–19° 44′	OC
7.8 m	⊕ 6′		VI	Difficult

Now for another of our observing challenges. This is a small, faint globular cluster with a barely resolved halo and even less resolved core. Nevertheless the cluster is worth a glance, at least once.

NGC 6626	Messier 28	18^h 24.5^m	–24° 52′	GC
6.8 m	⊕ 11.2′	IV		Moderate

Only seen as a small patch of faint light in binoculars, this is an impressive cluster in telescopes. With an aperture of 15 cm it shows a bright core with a few resolvable stars at the halo's rim. With a larger aperture the cluster becomes increasingly resolvable and presents a spectacular sight. It lies at a distance of about 22,000 light years, and the origin of the cluster poses a problem, as it does not seem to have many similarities to other clusters, in regard to the amount of "metals" it possesses. It is a metal-poor[5] cluster. Well worth seeking out for large-aperture telescope owners, as it is a lost gem.

NGC 6638	–	18^h 31.0^m	–25° 30′	GC
9.2 m	⊕ 7.0′		VI	Moderate

In moderate apertures this will appear as a small, featureless disc with a diffuse boundary. Even with a larger telescope, the halo will remain unresolved. Once seen, completely forgotten.

NGC 6637	Messier 69	18^h 31.4^m	–32° 21′	GC
7.6 m	⊕ 7.1′		V	Moderate

Visible as just a hazy spot in binoculars, it appears with a nearly star-like core in telescopes. Large aperture will be needed to resolve any detail and will show the myriad dark patches located within the cluster.

IC 4725	Collinder 382	18^h 31.6^m	–19° 15′	OC
4.6 m	⊕ 32′	40	I 3 m	Easy

---

[5]"Metal-poor" indicates it has low amounts of elements heavier than hydrogen and helium.

Visible to the naked eye, Messier 25 is a pleasing cluster suitable for binocular observation. It contains several star chains and is also noteworthy for small areas of dark nebulosity that seem to blanket out areas within the cluster, but you will need perfect conditions to appreciate this. Unique for two reasons: it is the only Messier object referenced in the *Index Catalogue (IC)*, and is one of the few clusters to contain a Cepheid-type variable star – U Sagittarii. The star displays a magnitude change from 6.3 to 7.1 over a period of 6 days and 18 h.

NGC 6645	Herschel VI-23	18^h 32.6^m	−16° 53′	OC
8.5 m	⊕ 15′	20	IV 1 m	Moderate

Somewhat of a challenge if you observe from an urban location, this is still a nice open cluster. Easily located under dark skies. In small to medium aperture telescopes it will appear as a fairly large but spread out cluster set against the background of the Milky Way.

Trumpler 33	Collinder 378	18^h 35.8^m	−32° 59′	OC/?
8.8 m	⊕ 3.5′		−	Difficult

This is a peculiar open cluster in so much as it doesn't seem to have been observed very much. This may be due to the fact that the few observations of it that do exist state name only a dozen or more stars scattered more or less randomly. Probably what we are looking at is nothing more than an asterism.

NGC 6656	Messier 22	18^h 36.4^m	−23° 54′	GC
5.1 m	⊕ 32′		VII	Easy

Wonderful! This is a truly spectacular globular cluster, visible under perfect conditions to the naked eye. Low-power eyepieces will show a hazy spot of light, while high power will resolve a few stars. A 15 cm telescope will give an amazing view of minute bright stars evenly spaced over a huge area. Often passed over by northern hemisphere observers owing to its low declination, it is actually quite close to us, at only 10,000 light years away, nearly twice as close as Messier 13.

NGC 6681	Messier 70	18^h 42.2^m	−32° 18′	GC
8.0 m	⊕ 7.8′		V	Moderate

A faint binocular object that is a twin of Messier 69. Best viewed with a large aperture, as with a small telescope, it is often mistaken for a galaxy. It lies at a distance of 35,000 light years.

NGC 6716	Collinder 394	18^h 54.6^m	−19° 53′	OC
6.9 m	⊕ 10′	20	IV 1 p	Moderate

This is a loose and spread-out cluster of a couple of dozen 7th magnitude stars. In medium to large apertures it will appear as if the cluster is split into two separate groups.

NGC 6715	Messier 54	18^h 55.1^m	−30° 29′	GC
7.6 m	⊕ 9.1′		III	Moderate

With telescopic apertures smaller than 35 cm the cluster remains unresolved and will show only a larger view similar to that seen in binoculars – a faint hazy patch of light. It has a colorful aspect – a pale blue outer region and pale yellow inner core. Recent research has found that the cluster was originally related to the Sagittarius dwarf galaxy, but that the gravitational attraction of our galaxy has pulled the globular from its parent. Among the globular clusters in the Messier catalog it is one of the densest as well as being the most distant.

NGC 6723	–	18^h 59.6^m	−36° 38′	GC
7.2 m	⊕ 11.0′		VII	Moderate

This is a nice globular cluster that, when observed in a medium aperture, will be fairly resolved with a 4′ halo surrounding a fairly concentrated core. Increasing both aperture and magnification should show an increased halo and a barely resolved core.

NGC 6809	Messier 55	19^h 40.0^m	−30° 22′	GC
6.3 m	⊕ 19′		XI	Easy

A lovely cluster that is easily seen in binoculars, and just visible with the naked eye. Small-aperture telescopes (15 cm) show a bright, easily resolved cluster with a nice concentrated halo. Because it is very open, a lot of detail can be seen such as star arcs and dark lanes, even with quite small telescopes. With a larger aperture, hundreds of stars are seen. It contains quite a large number of the mysterious blue stragglers.

NGC 6864	Messier 75	20^h 06.1^m	−21° 55′	GC
8.5 m	⊕ 6′		I	Moderate

A difficult object to locate with binoculars, as it is so small and faint. Even then it will only appear as a hazy spot (like so many others). Will show a bright core and a few resolved stars in the halo with 25 cm aperture telescopes. This is one of the most distant globular clusters in Messier's catalog, at 60,000 light years.

Notes

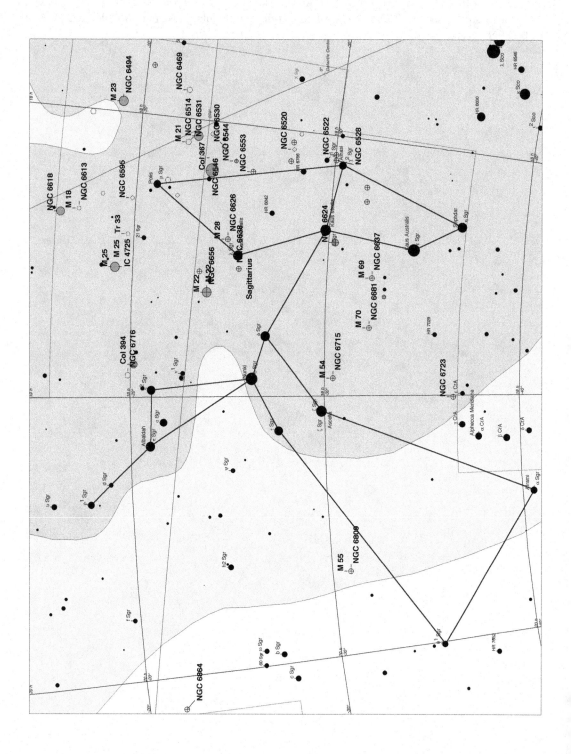

Notes (cont.)

Scorpius

Fast Facts

Abbreviation: Sco	Genitive: Scorpii	Translation: The Scorpion
Visible between latitudes 40° and −90°		Culmination: June

Clusters

NGC 6093	Messier 80	16ʰ 17.0ᵐ	−22° 59′	GC
7.9 m	⊕ 10.0′		II	Easy

Readily detectable in binoculars as a tiny, glowing, hazy patch set in a stunning star field, it has a distinctly noticeable brighter core. Telescopes will be needed to resolve its 14th magnitude stellar core. One of the few globular clusters to have been the origin of a nova, T Scorpii, when it flared to prominence in 1860, then disappeared back into obscurity within 3 months.

NGC 6121	Messier 4	16ʰ 23.6ᵐ	−26° 32′	GC
5.8 m	⊕ 26.3′		IX	Easy

Now for a superb object, presenting a spectacle in all optical instruments. It does however lie very close to the star, α Scorpii, or Antares, so that the glare may prove a problem in its detection with small aperture telescopes. High-power binoculars will even resolve several stars. Telescopes of all apertures show detail and structure within the cluster, and the use of high magnification will prove beneficial; but what is more noticeable is the bright lane of stars that runs through the cluster's center. Thought to be the closest globular to Earth at 7,200 light years (although NGC 6397 in Ara may be closer) and about 12.2 billion years old. Research has discovered many white dwarf stars in the cluster, believed to be among the oldest known, and one in particular is a binary star with a pulsar companion, PSR B1620-26, and a planet orbiting it with a mass 2.5 times that of Jupiter. It should be mentioned that there are a few unsubstantiated reports that the cluster has been glimpsed with the naked eye.

NGC 6124	Collinder 301	16ʰ 25.6ᵐ	−40° 40′	OC
5.8 m	⊕ 29′	75	I 3 r	Easy

A very nice rich cluster, suitable for large binoculars and small telescopes. There is a chain of stars at its southern edge, and a tightly grouped collection of five bright stars at its center. It also contains several nice star chains and a few red-tinted stars. It is relatively close, at a distance of around 1,500 light years. It is also known as Caldwell 75.

Collinder 302	–	16ʰ 26.1ᵐ	−26° 15′	OC
1.0 m	⊕ 505′	50	III 3 p	Easy

Also known as the Antares moving cluster, this little known object is very rarely observed. The reason for this is simple – it is enormous and encompasses most of the constellation. It is quite detached (an understatement), with no obvious concentration towards the center, and has such a large brightness range that it is almost impossible to see it as a cluster at all. It is also poorly populated considering its size.

NGC 6144	Herschel VI-10	16ʰ 27.2ᵐ	−26° 01′	GC
9.0 m	⊕ 9.3′		XI	Moderate

This nice globular suffers from being close to Antares, as it lies only 40′ northwest of it. But being so close does make it easy to locate. Light pollution might make it difficult to see, so medium apertures may be best, although give it a go with small apertures and see what transpires. When seen, however, its faint circular glow does contrast nicely with the ever-present red rays from its neighbor. High magnification may resolve its peripheral members.

NGC 6139	–	16ʰ 27.7ᵐ	−38° 51′	GC
9.1 m	⊕ 5.5′		II	Moderate

This globular cluster will appear faint and small in small to medium apertures, with an even brightness all the way to its center and no resolution. Higher magnification and larger aperture however will begin to resolve the cluster with a definite granularity becoming apparent. Under good seeing a handful of stars will seemingly blink in and out of view.

NGC 6178	Collinder 308	16ʰ 35.8ᵐ	−45° 38′	OC
7.2 m	⊕ 4′	12	III 3 p	Moderate

A rather non-descript open cluster containing over a dozen stars surrounding two 8th magnitude stars. Despite its dearth of bright stars, it does stand out rather well from the background with a distinctive triangular shape.

NGC 6192	Collinder 309	16^h 40.4^m	−43° 22′	OC
8.5 m	⊕ 7′	60	I2 r	Easy

With medium aperture this nice open cluster consists of about two to three dozen stars of 11th–14th magnitude set against an unresolved haze. Try using averted vision here, as many more will seem to become visible. Larger aperture will of course show more stars in pretty arcs and chains. This cluster is well worth seeking out and stands out well from the background star field.

NGC 6231	Collinder 315	16^h 54.0^m	−41° 48′	OC
2.6 m	⊕ 14′	100	I 3 p	Easy

A superb cluster also known as Caldwell 76 and located in an awe-inspiring region of the sky. Brighter by 2.5 magnitudes than its northern cousin the double cluster in Perseus. The cluster is full of spectacular stars: very hot and luminous O-type and B0-type giants and supergiants, a couple of Wolf-Rayet stars, and ξ^{-1} Scorpii, which is a B1.5 Ia extreme supergiant star with a luminosity nearly 280,000 times that of the Sun! The cluster is thought to be a member of the stellar association Sco OB1,[6] with an estimated age of 3 million years. A wonderful object in binoculars and telescopes, the cluster contains many blue, orange and yellow stars. It lies between μ^{1+2} Scorpii and ξ^{-1} Scorpii, an area rich in spectacular views. A good cluster to test the technique of averted vision, where many more stars will jump into view. Observe and enjoy.

NGC 6242	Collinder 317	16^h 55.5^m	−39° 27′	OC
6.4 m	⊕ 9′		I3 m	Easy

A very nice cluster in all apertures, although naturally, bigger will be better. About 20 stars will be seen in small telescopes with the cluster being rich and compact. Low power is best used here, as it will show most, if not the entire, cluster in the field of view. Using a medium aperture will show many more stars, nearly 50, with several of 8th and 9th magnitude set against many fainter stars. Larger aperture will reveal many star chains and arcs.

Collinder 316		16^h 55.5^m	−40° 50′	OC
6.6 m	⊕ 105′	40	I 2 m	Easy

---

[6] See entry at end of this section on Scorpius.

This is a large and sparse cluster with its stars loosely spread across 2°. This means it is often overlooked, which is a shame. In binoculars there are about three dozen of its stars ranging in brightness from 6th to 9th magnitudes. It can appear that the cluster mingles with Trumpler 24 (see below), and so it is difficult to say where one cluster ends and another begins.

Trumpler 24	Harvard 12	16ʰ 57.0ᵐ	−40° 40′	OC
8.6 m	⊕ 60′	100	IV 2 p n	Easy

A loose and scattered cluster, set against the backdrop of the Milky Way. Easily seen in binoculars where about four dozen stars can be seen, set against a spectacular background. Small apertures will show clumps and knots of stars, along with a couple of arcs and chains. It is, along with nearby Collinder 316, the core of the Scorpius OB1 stellar association. This whole region, including several of the clusters mentioned above, is swathed in very faint nebulosity, including the faint emission nebula, IC 4628, along with the nearby dark nebula Barnard 48. Don't use medium to large apertures here, as you will lose the spectacular impact of the cluster and its surroundings.

NGC 6249	Collinder 319	16ʰ 57.7ᵐ	−44° 48′	OC
8.2 m	⊕ 6′	30	II 2 m	Moderate

Consisting of over a dozen stars, ranging from 9th to 12th magnitude, this is a rather unimpressive open cluster. There isn't actually a lot more one can say except it's brightish and smallish.

NGC 6259	Collinder 322	17ʰ 00.7ᵐ	−44° 39′	OC
8.0 m	⊕ 10.0′	100	II 2 r	Easy

This is a lovely cluster that would be observed far more often if it were only further north in the sky. Nevertheless, when seen, it shows about one dozen stars set against an unresolved haze in medium apertures. In larger apertures it becomes splendid, with over a hundred stars resolved in a uniform circular cloud. There are several star arcs interspersed with dark lanes. Many observers report that it is similar in appearance to both Messier 11 in Scutum, and NGC 7789 in Cassiopeia. It really is a lovely object, but suffers from being near the horizon for observers in northern Europe and the northern United States.

NGC 6268	Collinder 323	17ʰ 02.2ᵐ	−39° 43′	OC
9.5 m	⊕ 6.0′	40	II 2 p	Moderate

This open cluster is distinctive due to the fact that it does not, as most clusters do, appear as an oval or circular group but more of a stream or an arc of stars. Although small and rich, it is bright, and in medium aperture will consists of around 20 stars ranging from 9th to 12th magnitude in two streams separated by a star free region. Larger aperture reveals more stars and chains, with outlying stars branching off.

Bochum 13	–	17^h 17.4^m	–35° 33′	OC
7.2 m	⊕ 14′	60	III 3 m	Easy

This is an open cluster that is not observed much, and reported even less. It is detached with little or no concentration towards the center with a large brightness range. There are a handful of reports that it can be glimpsed with the naked eye, but these are not substantiated.

NGC 6322	Collinder 326	17^h 18.4^m	–42° 56′	OC
6.0 m	⊕ 10.0′	30	I 3 m	Easy

Located in a very dusty part of the Milky Way, with its dark lanes, this open cluster is a nice triangular group of three 7th magnitude stars set among many more faint members. Larger apertures will reveal more cluster stars along with several pairs.

Collinder 332	–	17^h 30.8^m	–37° 05′	OC?
8.9 m	⊕ 10.0′	12	IV 1 p	Easy

We now have something of a dilemma. Try observing this and what do you see, just a patch of ordinary Milky Way or a definite open cluster? It may be nothing more than a slight enhancement of the background sky, and not a true cluster, but that's not a sure thing. The object is not well detached from the surrounding star field, has a small brightness range and is poorly populated.

NGC 6383	Collinder 335	17^h 34.8^m	–32° 34′	OC
5.5 m	⊕ 2.5′	40	IV 1 p	Easy

This is a nice object, but subject to much confusion. Is it NGC 6383, or NGC 6374, or Collinder 334, or Collinder 335? As far as I can ascertain, they are all the same object! If so, then it is a naked-eye object, appearing as a faint hazy blob north of the sting in the scorpion's tail. Small apertures will reveal over a dozen stars west of a 5th magnitude star. In larger apertures many more stars appear. The whole cluster in enveloped in very faint nebulosity.

Trumpler 27	Collinder 336	17^h 36.3^m	–33° 31′	OC
6.7	⊕ 6.0′	35	III 3 m	Moderate

This is a little group of stars that is sparse and open, like most Trumpler clusters. It may be difficult to discern the cluster members as only a dozen or so stand out, the rest being responsible for the granular background. From a research point of view, Trumpler 27 is a young open cluster of stars, at about 11 million years, that lies about 5° from the direction of the galactic center at a distance of 5,400 light years. Due to the interstellar dust lying between the cluster and us its visible light has been absorbed, making it been difficult to study.

NGC 6388	–	17ʰ 36.3ᵐ	–44° 44′	GC
6.8 m	⊕ 8.7′	–	III	Moderate

A medium or larger aperture is really needed for this globular cluster, and even then resolution of the core will not be achieved. However, a handful of stars may be resolved at its halo, and increasing the aperture will just show a granular surface. One redeeming feature is that its core has one of the highest surface brightnesses of any globular cluster, which is readily apparent.

Trumpler 28	–	17ʰ 37.0ᵐ	–32° 29′	OC
7.7 m	⊕ 6.0′	30	III 2 m n	Easy

Although a rather inconspicuous cluster, it is easy to locate, as it lies nearly halfway between Messier 6 and NGC 6383. It will appear detached and spread fairly evenly. Images show nebulosity, but whether this is within the cluster or in the line of sight is unknown at the moment.

NGC 6396	Collinder 339	17ʰ 37.6ᵐ	–35° 01′	OC
8.5 m	⊕ 3.0′	30	II 3 m	Moderate

This is a very small cluster that is not particularly impressive or rich, and subsequently will need at least a medium aperture to appreciate and a large aperture is preferable; nevertheless, it has about a dozen 10th magnitude stars in a small group. It does have two nice features, however; there is a distinctive chain of four or five stars running through the center, and there is a pleasant double star, h 4966, at the chain's end. Very large apertures will reveal that the double is in fact a quadruple!

Collinder 338	–	17ʰ 38.3ᵐ	–37° 43′	OC
8.0 m	⊕ 25.0′	40	III 2 m	Easy

This cluster can be glimpsed in binoculars as a patch of 8th magnitude stars set among an unresolved haze. By using averted vision several points of light will suddenly appear. Using even a small telescope the cluster tends to lose cohesiveness and will be lost among the background stars.

NGC 6405	Collinder 341	17ʰ 40.1ᵐ	−32° 13′	OC
4.2 m	⊕ 33′	100	II 3 r	Easy

This cluster, Messier 6, is easily seen with the naked eye as a dim patch of light. It is probably one of the few stellar objects that actually looks like the entity after which it is named, the Butterfly Cluster. This is American astronomer Phil Harrington's description of the cluster: "I know it's an illusion, but my 16×70 binoculars create a three-dimensional effect when I aim toward this pair of open clusters (including Messier 7). Many of the brighter cluster stars look as though they are floating in front of a field of fainter stardust. The effect really struck me about 10 years ago at the Stellafane amateur telescope making convention in Vermont, as I was watching M6 and M7 slowly graze along the tops of some distant pine trees. It had a kind of visual impact that could never be duplicated in a photograph." A fine sight in binoculars, it contains the lovely orange-tinted star BM Scorpii east of its center. This star is a semi-regular variable, period 850 days, which changes from magnitude 5.5 to 7. Surrounding it are many nice steely blue-white stars. Believed to be at a distance of 1,590 light years.

NGC 6400	Collinder 342	17ʰ 40.2ᵐ	−36° 56′	OC
8.8 m	⊕ 12.0′	60	II 2 m	Easy

This is a nice cluster that has a feature that is very apparent – many stars seem to radiate outward in lines and chains. With a medium aperture about two dozen stars will be seen, and use of a larger telescope will reveal fainter stars also in chains.

Trumpler 29	–	17ʰ 41.5ᵐ	−40° 09′	OC
6.7 p	⊕ 7.0′	30	III 2 m	Moderate

To really appreciate this cluster, a medium to large aperture is best. Then it will appear as a group of stars that is located within the twist of the scorpion's tail. Look for the starless void lying east of the cluster.

NGC 6416	Collinder 344	17ʰ 44.3ᵐ	−32° 21′	OC
5.7 m	⊕ 15.0′	40	III 2 m	Easy

This large and scattered cluster, of about thirty 8–11th magnitude stars, and many more faint stars, is not very impressive. Some star chains may be seen along with a couple of reddish stars along its apparent edge. It may be that the cluster extends further north, but the star density tends to meld into the background Milky Way.

NGC 6425	Collinder 344	17ʰ 47.0ᵐ	−31° 31′	OC
7.92 m	⊕ 15.0′	30	II 1 m	Easy

Another cluster that really needs a moderate to large aperture to be appreciated, this cluster shows a definite triangular shape, with several chains and arcs often separated by intriguing starless voids. It will seem as if its general shape is mirrored within the cluster, as several of the chains themselves contain triangular-shaped groups. Surprisingly for a cluster, it will show up well in a large aperture provided a wide field eyepiece is used.

NGC 6441	–	17ʰ 50.2ᵐ	−37° 03′	GC
7.2 m	⊕ 7.8′		III	Easy

After the admittedly lackluster clusters of the previous few entries, it is nice to come across a much more spectacular object. This globular cluster, visible even in a finderscope, when seen in a medium aperture will show a very bright but unresolved object with a quite intense core. Increasing the aperture will hint at granularity in the cluster. What can't be missed however is the bright 3rd magnitude star G Scorpii, with a lovely orange color, just 8′ west.

NGC 6475	Collinder 354	17ʰ 53.9ᵐ	−34° 49′	OC
3.3 m	⊕ 80′	80	I 3 r	Easy

This cluster, more usually called Messier 7, is an enormous and spectacular open cluster. It presents a fine spectacle in binoculars and telescopes, containing over 80 blue-white and pale yellow stars. It is only just over 800 light years away, but is over 200 million years old. Many of the stars are around 6th and 7th magnitude, and thus should be resolvable with the naked eye. Also known as Ptolemy's Cluster because it was first mentioned by Ptolemy, and if he could see it, so should you!

Trumpler 30	–	17ʰ 56.8ᵐ	−35° 16′	OC
8.8 p	⊕ 10.0′	20	IV 1 p	Moderate

This is a nice cluster, packed with very faint stars, so, inevitably, medium to large apertures will be needed. Then it will show a roughly triangular aspect with an 8th magnitude stars at its most northerly point. Increasing both aperture and magnification will reveal many more faint members. This cluster is somewhat of a challenge, but worth it.

NGC 6496	–	17^h 59.0^m	–44° 16′	GC
8.6 m	⊕ 6.9′		XII	Moderate

A difficult globular cluster to observe even from the southern United States, it will appear faint and small, set in a very rich star field. In large apertures some resolution will be on the threshold of your vision. It may appear that a few stars are resolved across the cluster, but it is believed that these are just foreground stars. It has however been the subject of much research. Although located towards the galactic bulge, the cluster is not a member of the subgroup of globular clusters that reside in or near the disc of the Milky Way, but rather, along with NGC 6624 and NGC 6637, halo clusters with strongly inclined orbits. Finally, some books place the cluster in Corona Australis, while others in Scorpius. Either way, it is very far south.

The Scorpius-Centaurus Association, 550 Light Years Away

A much older but closer association than the Orion Association. It includes most of the stars of 1st, 2nd and 3rd magnitude in Scorpius down through Lupus and Centaurus to Crux. Classed as a B-type association because it lacks O-type stars, its angular size on the sky is around 80°. It is estimated to be 750×300 light years in size, and 400 light years deep, with the center of the association midway between α Lupi and ζ Centauri. Its elongated shape is thought to be the result of rotational stresses induced by its rotation about the galactic center. Bright stars in this association include θ Ophiuchi, β, υ, δ, and σ Scorpii, α, γ Lupi, ε, δ, μ Centauri, and β Crucis.

Notes

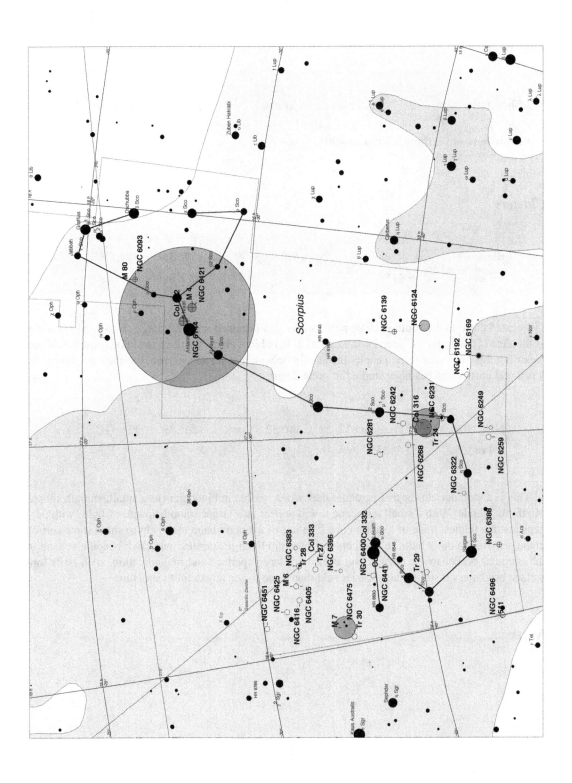

Sculptor

Fast Facts

Abbreviation: Scl	**Genitive: Sculptoris**	**Translation: The Sculptor's Workshop**
Visible between latitudes 50° and –90°		**Culmination: September**

Clusters

Blanco 1	**–**	**00ʰ 04.3ᵐ**	**–29° 56′**	**OC**
4.5 m	**⊕ 90′**	**30**	**III 2 m**	**Easy**

Located close to the south galactic pole, this is an ill-defined and very large cluster. Also known as the Zeta (ζ) Sculptoris Cluster, it is easily visible in binoculars. There seems to be some confusion regarding this object, as some people mistakenly believe it to be an asterism, but they are wrong, as it is a real cluster and has been studied in great detail.

NGC 288	**Herschel VI-20**	**00ʰ 52.8ᵐ**	**–26° 35′**	**GC**
9.4 m	**⊕ 13.8′**		**X**	**Easy**

This is a globular cluster that is, under dark skies, visible in binoculars as a small, smooth sphere of 8th magnitude. With a small telescope it will appear as a larger smooth sphere of light, with perhaps just the faintest trace of a dense core. The use of averted vision may help to show some sort of granularity to the outer halo. In telescopes of medium to large aperture, what will become evident is the irregular border to the cluster. But even with large aperture and magnification, such is its low surface brightness it still remains easier said than done to see any definite structure.

Notes

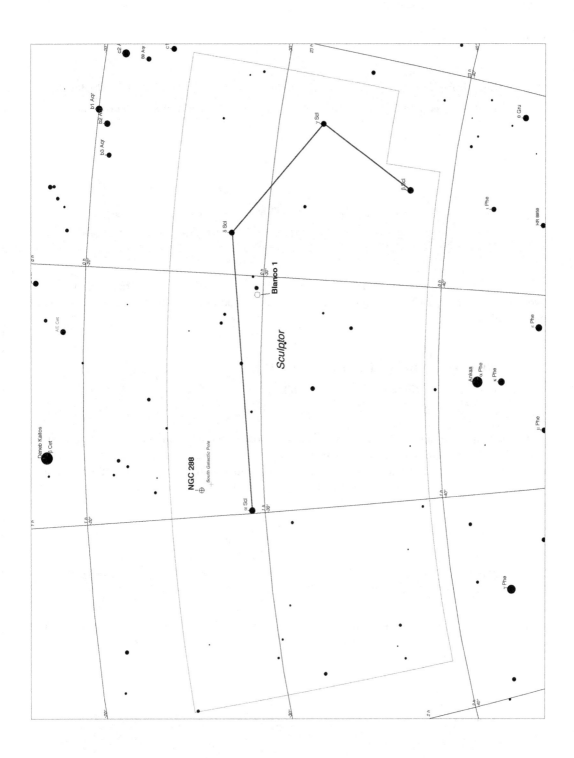

Scutum

Fast Facts

Abbreviation: Sct	**Genitive: Scuti**	**Translation: The Shield**
Visible between latitudes 70° and −90°		**Culmination: June/July**

Clusters

NGC 6649	Collinder 384	18ʰ 33.4ᵐ	−10° 24′	OC
8.9 m	⊕ 5.0′	30	I 3 m	Moderate

Set in a particularly dark field, due to a plethora of dark nebulae, this is a fairly rich cluster of about two dozen stars around 12th magnitude and fainter. Larger aperture reveals more stars set among an unresolved haze.

NGC 6664	Herschel VIII-12	18ʰ 36.5ᵐ	−08° 11′	OC
7.8 m	⊕ 12.0′	50	III 2 m	Moderate

This is a large but faint scattered open cluster that is overpowered by the ever-present glare of the 4th magnitude star alpha (α) Scuti. It can be a difficult object to locate under an urban sky, but with the right conditions, can be located in binoculars. With increasing apertures the number of stars visible increases, as expected, along with a few star chains, but overall it isn't a particularly impressive object.

Trumpler 34	–	18ʰ 39.8ᵐ	−08° 25′	OC
8.6 m	⊕ 7.0′	40	II 2 m	Difficult

We now have a small cluster that may be a challenge for small apertures, or poor conditions. With appropriate aperture and under a decent sky, you will see a small group of around two dozen stars of 13th–15th magnitude. Although not visually impressive, it is a good test of both your optics and sky conditions.

NGC 6683	–	18ʰ 42.2ᵐ	−06° 12′	OC
9.4 m	⊕ 11.0′	20	II I p n	Moderate

The problem we have here isn't so much due to the faintness of the cluster and dearth of stars, but rather that it lies on the edge of the wonderful Scutum Star Cloud and the Great Rift, and this tends to overwhelm the cluster. It doesn't stand out particularly well from the surrounding field, but a moderate aperture will show about a dozen faint stars in a roughly elongated group over an unresolved background haze. One cannot but be amazed by the star fields and dark clouds that the cluster is immersed in. Overall this is an outstanding part of the sky. One could truly loose oneself in this part of the sky, scanning it for several hours and never being bored.

NGC 6694	Collinder 389	18^h 45.2^m	−09° 24′	OC
8.0 m	⊕ 14′	30	I 2 m	Moderate

Also known as Messier 26, this is a small but rich cluster containing 11th and 12th magnitude stars, set against a haze of unresolved stars. This makes it unsuitable for binoculars, as it will be only a hazy small patch of light, and so apertures of 10 cm and more will be needed to appreciate it in any detail.

NGC 6704	Collinder 390	18^h 50.7^m	−05° 12′	OC
9.2 m	⊕ 5.0′	30	I 2 m	Difficult

A small and faint cluster, containing about two dozen stars set against an unresolved haze. Even with a large aperture the only gain is the number of stars seen. However, the attraction of this cluster isn't so much its stars but the nearby lovely reddish star that is 2′ to the northwest.

NGC 6705	Collinder 391	18^h 51.1^m	−06° 16′	OC
5.8 m	⊕ 13′	200	I 2 r	Moderate

Also known as Messier 11, with the popular name the Wild Duck Cluster, this is a gem of an object. Although it is visible with binoculars as a small, tightly compact group, reminiscent of a globular cluster, they do not do it justice. With telescopes, however, its full majesty becomes apparent. Containing many hundreds of stars, it is a very impressive cluster. It takes high magnification well, where many more of its 700 members become visible. At the top of the cluster is a glorious pale yellow tinted star. The British amateur astronomer Michael Hurrell calls this "one of the most impressive and beautiful celestial objects in the entire sky." This is my personal favorite of all open clusters.

NGC 6712	Herschel I-47	18^h 53.1^m	−08° 42′	GC
8.3 m	⊕ 9.8′		IX	Easy

This is a fairly large and moderately bright globular cluster set in a rich field. It is easily found in binoculars, and with a small aperture telescope partial resolution can be achieved; however, it is difficult to see where the cluster ends and the background star field begins. Larger apertures will start to show some granularity. It lies within the same field as the planetary nebula IC 1295.

Notes

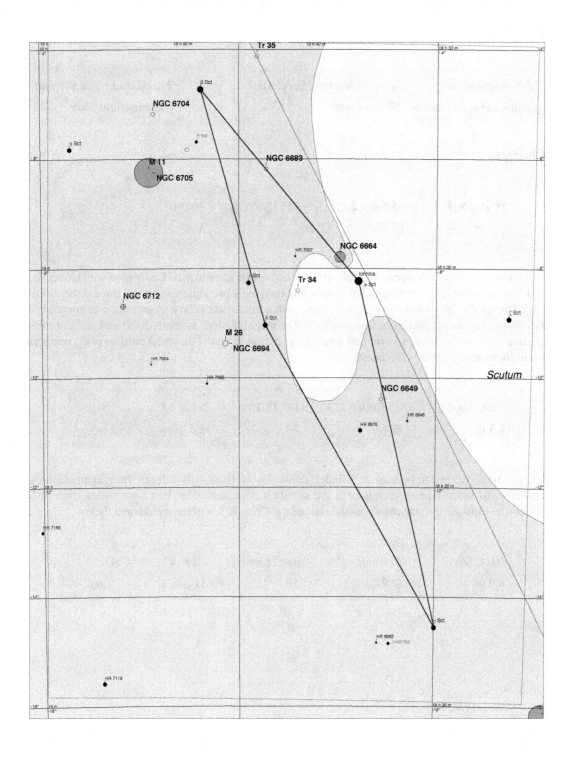

Serpens[7]

Fast Facts

Abbreviation: Ser	Genitive: Serpentis	Translation: The Serpent
Visible between latitudes 80° and –80°		Culmination: May

Clusters

NGC 5904	Messier 5	15ʰ 18.6ᵐ	+02° 05′	GC
5.7 m	⊕ 23.0′		V	Easy

This is a wonderful cluster and visible to the naked eye on clear nights. Easily seen with binoculars as a disc, and with large telescopes the view is breathtaking – presenting an almost three-dimensional vista. One of the few colored globular clusters, with a faint, pale yellow outer region surrounding a blue-tinted interior. It gets even better with higher magnification, as more detail and stars become apparent. Possibly containing over half a million stars, this is one of the finest clusters in the northern hemisphere; many say it is *the* finest.

NGC 6604	Collinder 373	18ʰ 18.2 ᵐ	–12° 14′	OC
6.5 m	⊕ 4.0′	30	I 3 m n	Easy

This open cluster can be seen in a finder, consisting of about half a dozen 7th magnitude stars. However, its most impressive feature is the nebula it is immersed in, best seen with a filter. It is difficult to estimate the star colors, as the dust of the Great Rift will have reddened them.

NGC 6611	Collinder 375	18ʰ 18.8ᵐ	–13° 47′	OC
6.0 m	⊕ 22′	50	II 3 m n	Easy

---

[7] Serpens actually consists of two constellations, Serpens Caput and Serpens Cauda, separated by Ophiuchus.

We now have a fine large cluster easily seen with binoculars. It is about 7,000 light years away, located in the Sagittarius-Carina spiral arm of the galaxy. Its hot O-type stars provide the energy for the Eagle Nebula, within which the cluster is embedded. A very young cluster of only 800,000 years with a few stars at 50,000 years old. Also known as Messier 16, which is the name for both the cluster and nebula.

IC 4756	Collinder 386	18^h 39.0^m	+5° 27′	OC
4.6 m	⊕ 52′	80	II 3 r	Easy

An enormous cluster often missed by observers, consisting of nearly six dozen 7–11th magnitude stars. Spread over a nearly a degree of sky, it is a splendid sight in large binoculars or small aperture telescopes with several small groups of stars set among the larger collection, but larger apertures tend to lose the clustering effect. There are a few unsubstantiated reports that it can be seen with the naked eye. Give it a go and find out for yourself.

Notes

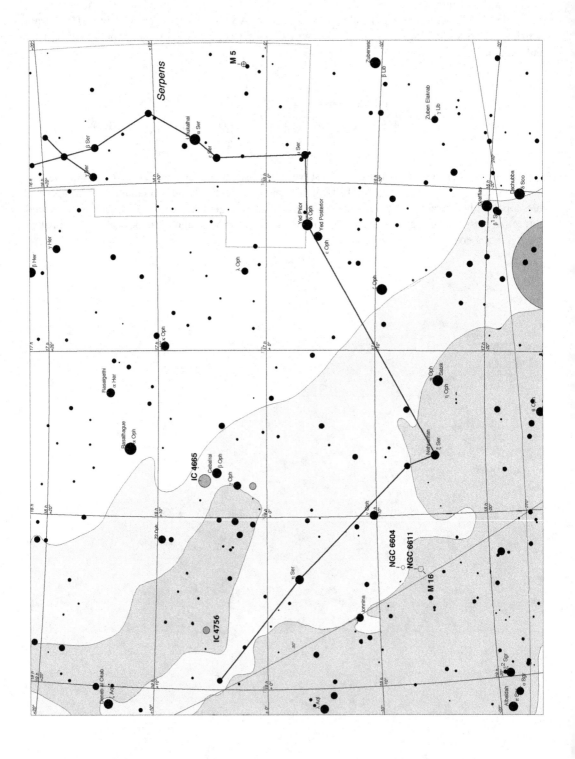

Notes (cont.)

Taurus

Abbreviation: Tau	Genitive: Tauri	Translation: The Bull
Visible between latitudes 90° and −65°		Culmination: November/ December

Melotte 22	Messier 45	03ʰ 47.4ᵐ	+24° 07′	OC
1.6 m	⊕ 110′	100	I 3 r	Easy

Outstanding! Without a doubt the sky's premier star cluster. The Seven Sisters, or Pleiades, is beautiful however you observe it – naked eye, through binoculars or with a telescope. To see all the members at one go will require binoculars or a rich field telescope. Consisting of over 100 stars, spanning an area four times that of the full Moon, it will never cease to amaze. It is often stated that from an urban location 6 to 7 stars may be glimpsed with the naked eye. However, it may come as a surprise to many of you that it has 10 stars brighter than 6th magnitude, and that seasoned amateurs with perfect conditions have reported 18 being visible with the naked eye. It lies at a distance of 410 light years away, is about 115 million years old, and is the fourth-nearest cluster. It contains many stunning blue and white B-type giants. The cluster contains many double and multiple stars. Under perfect conditions with exceptionally clean optics, the faint nebula NGC 1435, the Merope Nebula surrounding the star of the same name (Merope – 23 Tauri), can be glimpsed and was described by W. Tempel in 1859 as "a breath on a mirror." However, this and the nebulosity associated with the other Pleiades are not, as they were once thought, to be the remnants of the original progenitor dust and gas cloud. The cluster is just passing through an edge of the Taurus Dark Cloud Complex. It is moving through space at a velocity of about 40 km a second, so by A.D. 32,000 it will have moved an angular distance equal to that of the full Moon. The cluster contains the stars Pleione, Atlas, Alcyone, Merope, Maia, Electra, Celaeno, Taygeta and Asterope. What we have here is a true celestial showpiece.

Melotte 25	Caldwell 41	04ʰ 27.0ᵐ	+15° 52′	OC
0.5 m	⊕ 330′	40	II 3 m	Easy

Also known as the Hyades. The nearest cluster after the Ursa Major Moving Stream, lying at a distance of 151 light years, with an age of about 625 million years. Even though the cluster is widely dispersed both in space and over the sky, it nevertheless is gravitationally bound, with the more massive stars lying at the center of the cluster. Best seen with binoculars, owing to the large extent of the cluster – over 5½°. Hundreds of stars are visible, including the fine orange giant stars γ, δ, ε and θ⁻¹. Tauri. Aldebaran, the lovely orange K-type giant star, is not a true member of the cluster but is a foreground star only 70 light years away. Visible even from light polluted urban areas – a rarity!

NGC 1647	–	04ʰ 45.9ᵐ	+19° 06′	OC
6.4 m	⊕ 45′	50	II 2r	Easy

This is a large and irregularly scattered cluster that can be glimpsed with binoculars and averted vision will reveal many more. Using larger and larger apertures will naturally reveal many more of the 11th and 12th magnitude stars. It contains many apparent double and triple star systems.

NGC 1746	Melotte 28	05^h 03.6^m	+23° 49′	OC/Ast
6.1 m	⊕ 42′	20	III 2 p	Easy

Another large and scattered cluster, visible on clear nights with the naked-eye. Current research suggests that in fact NGC 1746 is not a true cluster but an asterism, and that what we see is rather an optical grouping of stars. Actually lying behind the cluster are two other smaller clusters, each with its own classification – NGC 1750 and 1758.

NGC 1807	Melotte 29	05^h 10.7^m	+16° 32′	OC
7.0 m	⊕ 17′	20	II 2 p	Moderate

This is a fairly loose and irregular open cluster where averted vision can show many more of its fainter members. It contains over 30 9th magnitude stars, all visible in a 12′ field of view. There is some suggestion that it may be no more than an asterism, but this is not certain.

NGC 1817	–	05^h 12.2^m	+16° 41′	OC
7.7 m	⊕ 16′	40	IV 2 r	Easy

Lying just northeast of NGC 1807, this cluster can be glimpsed in binoculars. In fact, both clusters can be seen together in the same field of view, and thus has led to it being given the name (somewhat sarcastically) as the Poor Man's Double Cluster. The cluster is fainter and less impressive than its companion, but worth a look nevertheless.

Collinder 65	–	05^h 25.4^m	+16° 06′	OC 6
3.0 m	⊕ 220′	50	II 3 p	Easy

This is an enormous naked-eye open cluster that lies across the borders of both Taurus and Orion. Best seen in binoculars, it has about six members that are brighter than 6th magnitude, with some unsubstantiated reports of nebulosity surrounding the star 116 Tauri. It is a very nice object to observe with a bright orange star at its northern edge, 119 Tauri. Oddly, there is not much information on this cluster, and research suggests it may be an OB Association.

The Hyades Stream

There is some evidence (although it is not fully agreed upon) that the Ursa Major Stream is itself located within a much older and larger stream. This older component includes Messier 44, Praesepe in Cancer, and the Hyades in Taurus, with these two open clusters being the core of a very large but loose grouping of stars. Included within this are Capella (α Aurigae), α Canum Venaticorum,[8] δ Cassiopeiae and λ Ursae Majoris. The stream extends to over 200 light years beyond the Hyades star cluster, and 300 light years behind the Sun. Thus, the Sun is believed to lie within this stream.[9]

Notes

[8] Capella and a Canum Venaticorum are also thought to be members of the even larger Taurus Stream, which has a motion through space similar to the Hyades, and thus may be related.

[9] The bright stars that extend from Perseus, Taurus and Orion, and down to Centaurus and Scorpius, including the Orion and Scorpius-Centaurus associations, lie at an angle of about 1.5° to the Milky Way, and thus to the equatorial plane of the galaxy. This group or band of stars is often called Gould's Belt.

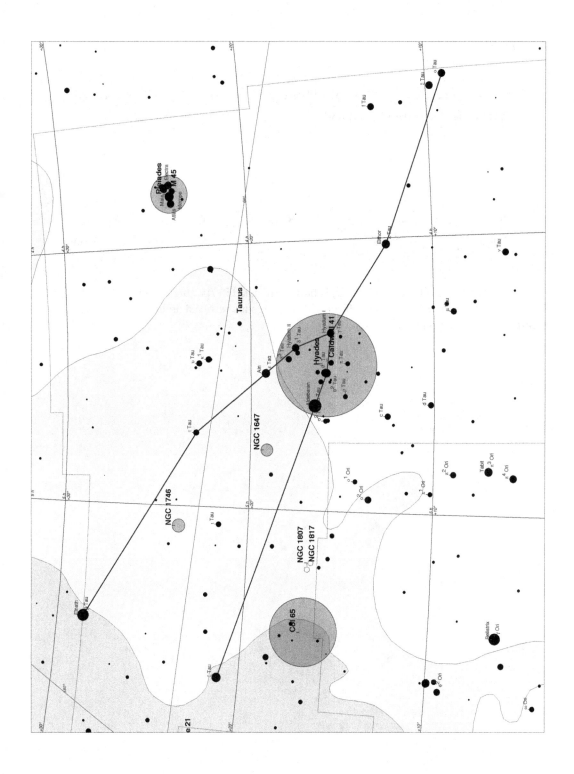

Telescopium

Fast Facts

Abbreviation: Tel	**Genitive: Telescopii**	**Translation: The Telescope**
Visible between latitudes 30° and −90°		**Culmination: July**

Clusters

NGC 6584	–	18^h 18.6^m	−52° 13′	**GC**
7.9 m	⊕ 6.6′		**VIII**	**Difficult**

Telescopium's sole cluster, a globular, is best appreciated in apertures around 30 cm and greater. Then it will be observed as a bright object with several stars resolved around its edge. It has a nice, well-formed, and concentrated center.

Notes

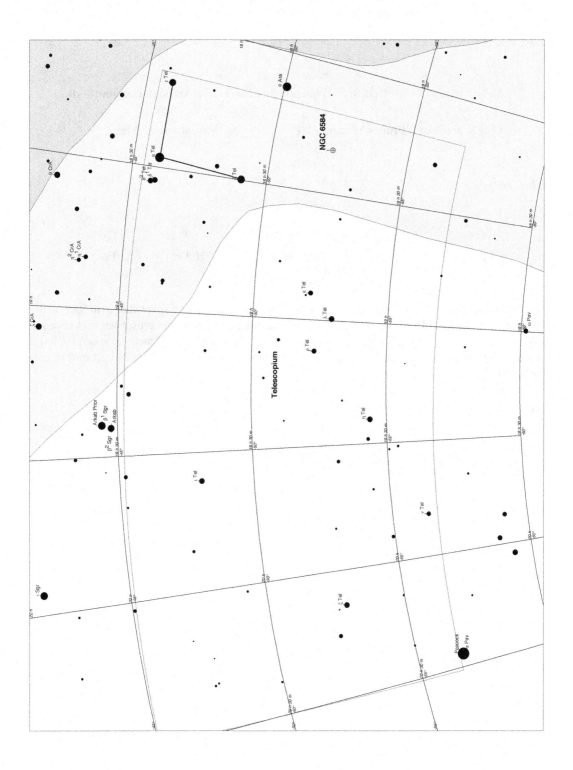

Triangulum Australe

Fast Facts

Abbreviation: TrA	Genitive: Trianguli Australe	Translation: The Southern Triangle
Visible between latitudes 15° and −90°		Culmination: May

Cluster

NGC 6025	–	16ʰ 03.3ᵐ	−60° 25′	GC
5.1 m	⊕ 15′	100	II 3 r	Easy

This is a nice open cluster that can be glimpsed in a finderscope as a somewhat oval haze with a brightish star at one end and is especially good in binoculars. In small apertures it begins to become impressive, appearing as a rich, detached cluster. It really starts to become something special in larger apertures, with many more stars being resolved, looping in arcs and chains set against a dark background.

Notes

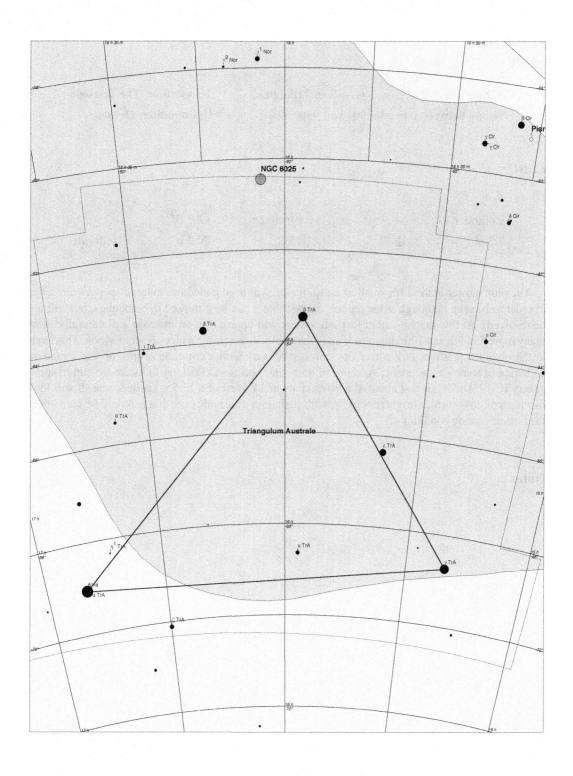

Triangulum

Fast Facts

Abbreviation: Tri	Genitive: Trianguli	Translation: The Triangle
Visible between latitudes 90° and −50°		Culmination: October

Cluster

Collinder 21	–	01ʰ 50.2ᵐ	+27° 05′	OC
7.3 m	⊕ 7′	15	IV 2 p	Moderate

The only cluster in the 11th smallest constellation will need a medium to large aperture telescope in order to be seen (although some reports suggest that it can be glimpsed in binoculars, but will not be resolved). At low magnification just a few stars will appear, but an increase will naturally show many more 8th, 9th and 10th magnitude stars that really do resemble a nice crescent Moon. This leads to the obvious question. Is it a real open cluster of stars with a common origin, or just a random gathering of stars – an asterism? As a test of both your optics and skill, try to locate the barred spiral galaxy IC 1731 that lies just about 4′ north and about 14′ west of a 13.5 magnitude star. It will look like a small, low-surface brightness glow, with perhaps a miniscule core that can only be seen under the best observing conditions.

Notes

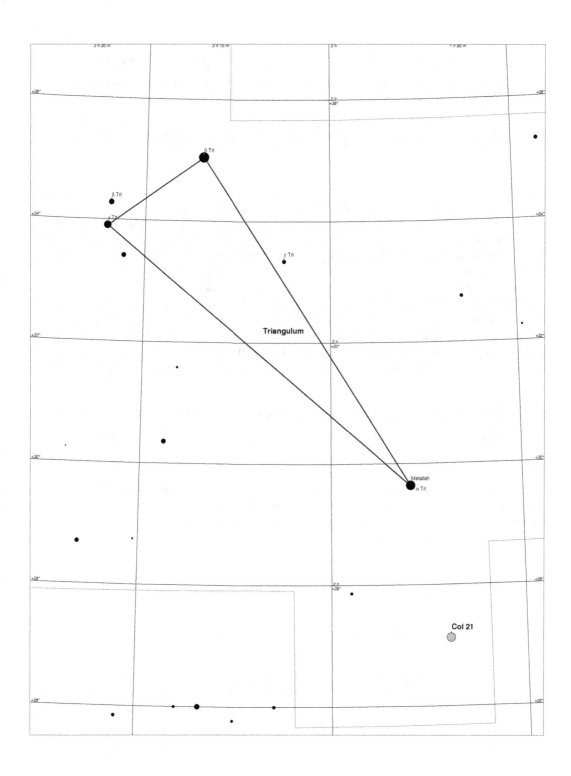

Tucana

Abbreviation: Tuc	Genitive: Tucanae	Translation: The Toucan
Visible between latitudes 15° and −90°		Culmination: September

NGC 104	47 Tucanae/Caldwell 106	00ʰ 24.1ᵐ	−72° 05′	GC
4.0 m	⊕ 50′		III	Easy

There may only be a few clusters in Tucana, but one of them is breathtaking! Regarded as the second finest cluster in the entire sky, 47 Tucanae is a showpiece object. Using the naked eye, the cluster will appear as a 4th magnitude out of focus star, and in binoculars it offers a wonderful vista of a star-like nucleus surrounded by a hazy glow. Increasing aperture only increases its magnificence. Seemingly thousands of tiny points of light spread out from its core. Some observers report that delicate and faint star chains can be seen, as well as the cluster having a faint yellow hue. Words really cannot convey the magnificence of this object. It is one of those objects that every observer should see at least once in his or her lifetime. As an added bonus, about 10′ from the apparent edge of the cluster's halo is the always overlooked globular, NGC 121. It is not surprising actually that it is infrequently observed, as it's a feeble 11.3 magnitude and tiny. NGC 121's one claim to fame is that blue stragglers were discovered in it, the first ever in an extragalactic globular cluster.

NGC 330	−4	00ʰ 56.3ᵐ	−72° 28′	OC
9.6 m	⊕ 1.4′	−	−	Easy

Lying within the Small Magellanic Cloud, this is one of the few objects lying in the cloud that stands out. It can be glimpsed in a finderscope and binoculars, displaying a nebula-like glow. In small apertures it will still seem nebulous, but bright and round. In moderate and larger apertures it becomes quite spectacular with a definite granularity.

NGC 362	Caldwell 104	01ʰ 03.2ᵐ	−70° 51′	GC
6.4 m	⊕ 14′		III	Easy

A naked-eye cluster, under dark skies of course, in binoculars it will look like an out-of-focus star. In moderate aperture telescopes, say 20 cm, it appears as a small spherical ball of unresolved light. Larger aperture will of course resolve more of the cluster, but there is some debate among observers as to whether it can be fully resolved with increasing aperture and/or magnification, or mostly resolved. In time the issue will probably be decided one way or another.

Notes

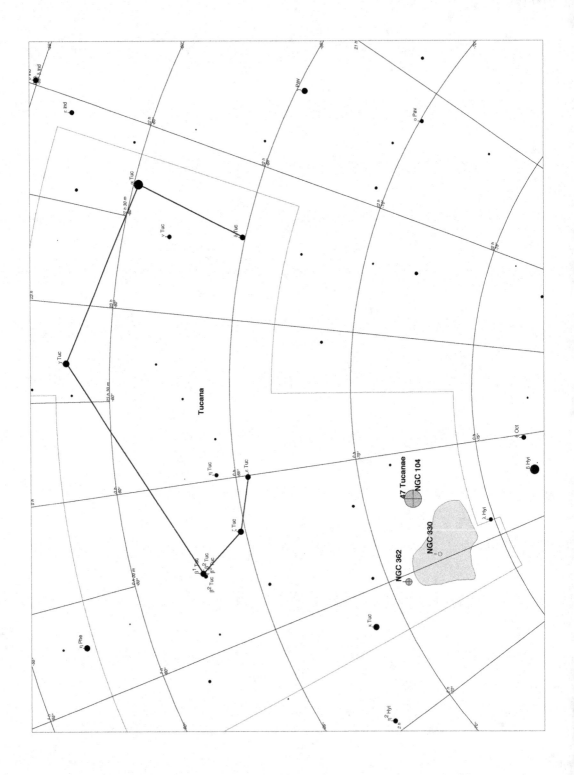

Notes (cont.)

Ursa Major

Fast Facts

Abbreviation: UMa Genitive: Ursae Majoris	**Translation: The Greater Bear**
Visible between latitudes 90° and −30°	**Culmination: March**

Cluster

The Ursa Major Stream

OK, this is going to seem a tad strange, but technically there are no clusters, either open or globular, that can be seen in Ursa Major. However, if we stretch the rules a bit we can include the truly enormous Ursa Major Stream that contains the five central stars of the Big Dipper or Plough. It is spread over a vast area of the sky, approximately 24°, and is around 20×30 light years in extent. It includes as members Sirius (α Canis Majoris), α Coronae Borealis, δ Leonis, β Eridani, δ Aquarii and β Serpentis. Due to the predominance of A1 and A0 stars within the association, its age has been estimated at 300 million years. Continuing in the vein of stretching things, we could say (with tongue firmly in cheek) that the cluster, or stream, is the closest open cluster to us, with its center only 70 light years distant. Nevertheless, this is always a good topic to present to unsuspecting members of the public at observatory open nights and star parties.

Notes

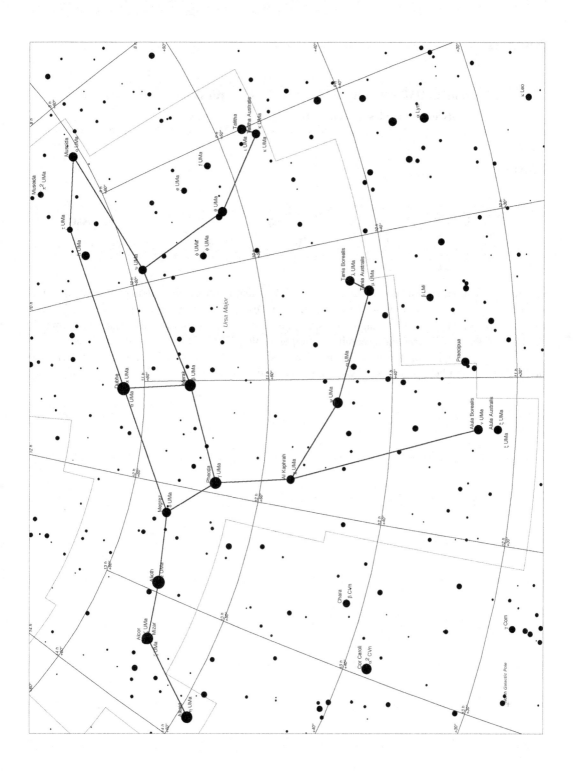

Ursa Minor

Fast Facts

Abbreviation: UMi	**Genitive: Ursae Minoris**	**Translation: The Lesser Bear**
Visible between latitudes 90° and –10°		**Culmination: May**

Clusters

NGC 5385	–	**13^h 52.4^m**	**+76° 11′**	**Open/Ast**
–m	⊕ 10′	10	–	**Difficult**

One of the "nonexistent" open clusters in the *RNGC* listing, this is a challenge. You will need an aperture of at least 40 cm to glimpse this grouping, and even then it will only appear as a well spread out ensemble of 11th and 12th magnitude stars. There is an 8th magnitude field star approximately 15′ ENE of the cluster. Recent research has suggested that NGC 5385 is in fact not a cluster at all, just a random grouping of non-related stars, and so should properly be referred to as an asterism. If nothing else it will be a good test of your telescope optics.

Notes

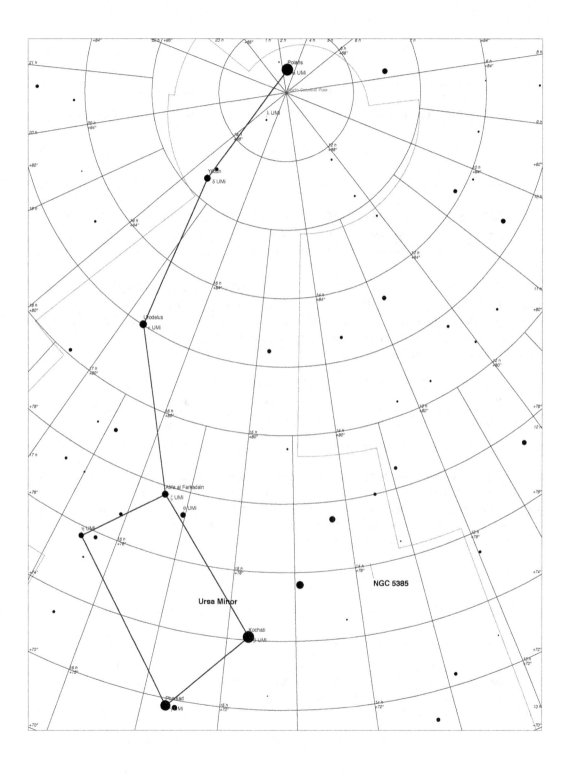

Vela

Fast Facts

Abbreviation: Vel	Genitive: Velorum	Translation: The Sail
Visible between latitudes 30° and −90°		Culmination: February

Clusters

NGC 2547	Collinder 177	08 ^{h} 10.2^m	−49° 14′	OC
4.7 m	⊕ 20′	100	I 3 r n	Easy

An open cluster that can be glimpsed with the naked eye, this is a splendid object in binoculars as well as in telescopes. It has over 100 members with brightnesses from 7th to 11th magnitude, and the periphery of the cluster merges into the background. One nice feature in particular is the distinctive curving arc of stars to the east along with a 6.5 magnitude star along its path. Larger apertures will reveal more arcs and streams.

Pismis 4	–	08^h 34.6^m	−44° 18′	OC
5.9 m	⊕ 25′	45	III 3 m n	Easy

This is an open cluster located at the edge of the Vela supernova remnant, consisting of a narrow stream of about two dozen various magnitude stars. There is a nice 7th magnitude orange star at its northern end. The nebula itself will need a filter in order to be glimpsed. Paris Marie Pismis was an Armenian astronomer who studied open clusters and was the first woman to graduate from the scientific institution of Istanbul University.

IC 2391	Collinder 191	08^h 40.3 ^{m}	−52° 53′	OC
2.5 m	⊕ 50′	45	III 3 m n	Easy

This is an enormous open cluster, or supercluster to be specific, that can be seen with the naked eye. Also known as the Omicron Velorum cluster and Caldwell 85, it is splendid to look at in binoculars, when about a dozen bright stars will be visible along with many fainter cluster members. There are some nice colored stars in the cluster, including its namesake, which has a delicate blue hue. Note that at its eastern edge is a small group of stars that could be taken as a cluster within a cluster. Some observers state that the cluster loses its appeal in anything bigger than a small aperture, whereas others disagree. What do you think?

IC 2395	–	08^h 42.5^m	−48° 06′	OC
4.6 m	⊕ 13′	40	III 3 m	Easy

A cluster that is easily visible in binoculars with over three dozen stars becoming visible in larger apertures, it will appear as a curving stream of stars with a bright star at one end of the arc. Not a stunning cluster by any means but still worth observing in my humble opinion.

NGC 2660	Collinder 193	08^h 42.6^m	−47° 12′	OC
8.8 m	⊕ 4.0′	70	I 1 r	Moderate

A small and faint open cluster in small apertures, it will just show as a hazy glow sprinkled with a handful of stars, somewhat similar to a globular cluster. However, it does improve somewhat in medium to large apertures, where it will stand out well against the background field. Increasing both aperture and magnification will reveal nearly four dozen 13th and 14th magnitude stars.

Collinder 197	–	08^h 44.8^m	−41° 22′	OC
6.7 m	⊕ 17′	40	III 3 m n	Easy

This is a large but irregularly scattered group of about 40 stars that has a nice double star at its center. There are a handful of unsubstantiated reports that this is a binocular object, but nothing confirmed. What is interesting is that the cluster is immersed in a faint nebula, Gum 15, that can be seen in a telescope, but will, of course, look better with the use of a filter.

Bochum 7		08^h 44.8^m	−45° 58′	OC
6.8 m	⊕ 20′	12	IV 3 p	Moderate

This is a cluster that is often overlooked, and it's easy to understand why. It has only about a dozen stars in a loose and irregular aspect that do not stand out at all when seen against the rich background. So it could be a challenge under an urban sky. Nevertheless it would be a good test of your observing skill and telescope optics.

NGC 2670	Collinder 200	08^h 45.5^m	−48° 48′	OC
7.8 m	⊕ 9′	30	III 2 m	Easy

After the somewhat disappointing preceding cluster we now have a rather fine open cluster that has a distinctive shape. Some observers describe it as a triangle, while others say it is more like a "Y" shape. It is bright, so should not be a problem to locate. It is visible in binoculars, but will not be as distinctive.

NGC 2669	Collinder 202	08ʰ 46.4ᵐ	−52° 56′	OC
6.1 m	⊕ 12′	90	III 3 m	Easy

A little known cluster that has a about a dozen fairly bright stars visible in small aperture, but considerably more faint with a medium and large aperture. However, once again there is some debate in the community as to whether it is a naked-eye object.

Trumpler 10	Collinder 203	08ʰ 47.7ᵐ	−42° 31′	OC/Ass
4.6 m	⊕ 29′	40	II 3 m	Easy

Now for something that is rather nice. This open cluster can be glimpsed in a finderscope, and is easily seen in binoculars from an urban location. It is very large, nearly the same size as the Moon, and is a sprawling group of bright and faint stars that fade at its edge into the background star field. It has been called the Poor Man's Pleiades, but that may be just a bit over the top. Research suggests that it is not a cluster, but rather an OB Association situated near the morphological center of the Gum Nebula and is in fact responsible for making most of the nebula glow, or fluoresce, along with the association Vela OB2.

IC 2488	Collinder 208	09ʰ 27.6ᵐ	−57° 00′	OC
7.4 p	⊕ 14′	50	II 3 r	Easy

Despite being a fairly bright open cluster, this is an oft-overlooked object. It can be glimpsed in binoculars as a small hazy spot, and in a small telescope about three dozen stars become visible. Look out for the parallel lines of stars in its center. Larger apertures will of course show many more of its 70 members. Those individuals blessed with large equipment may be able to see the planetary nebulae RCW 44 nearly one arc minute northeast of the cluster's center. Recent research suggests that the nebula is not associated with the cluster.

NGC 2910	Collinder 209	09ʰ 30.5ᵐ	−52° 55′	OC
7.2 m	⊕ 5′	60	III 3 m	Easy

This is a small group of 9th and fainter stars in an oval formation. With a medium aperture and low magnification the cluster will look more like a concentration of field stars, only becoming a self contained cluster with an increase of magnification. The whole cluster is set against an unresolved background haze of stars.

NGC 2972 – NGC 2999	Collinder 211	09ʰ 40.2ᵐ	–50° 19'	OC
9.9 m	⊕ 4'	60	II I p	Moderate

This cluster is included for its historical aspect. It appears that NGC 2972 is in fact the same object as NGC 2999. Best seen in medium and large apertures, where it will appear as a nice cluster that stands out against the background field.

Collinder 213	–	09ʰ 54.7ᵐ	–50° 55'	OC/?
9.2 m	⊕ 17'	–	–	Moderate

This object is not observed very often, and the reason is simple. It is faint, loose, and there is considerable debate as to whether it is a cluster at all. Two schools of thought exist: it is a cluster but not a very impressive one, or it is just a slight concentration of background stars. Whatever its true nature, it nevertheless can be a difficult object to locate with small apertures when, under perfect conditions it may be glimpsed as a faint and hazy spot. Larger apertures just show a scattering of stars. Observe for yourself and decide what it is.

NGC 3105	Collinder 214	10ʰ 00.6ᵐ	–54° 47'	OC
9.7 m	⊕ 2'	90	I 2 p	Easy

Now for an observing challenge, this scientifically interesting open cluster stands out for the simple reason that the background is faint and sparse. Of course, when we say it stands out, we are referring to the fact that you are using a medium to large aperture telescope because this is a very small object. When eventually found, it will appear as nothing more than a unresolved hazy cloud with just a handful of resolved 12th and 13th magnitude stars. Not exactly heart stopping, but nevertheless a nice object for those of you that like to test both your observing skill and telescope optics. From a research viewpoint, the cluster is very young and very distant.

NGC 3201	Caldwell 79	10ʰ 17.6ᵐ	–46° 25'	GC
6.7 m	⊕ 18'		X	Easy

This is a remarkable globular cluster in every respect. It is, barely, a naked-eye object; it can be glimpsed in a finderscope and starts to show its magnificence in binoculars. Small apertures will show a slightly oval-like shaped cluster with a hint of structure. By increasing magnification much more is resolved with stars seemingly branching off in all directions. In fact, under all apertures and magnifications this will be a splendid object and will repay long and repeated observation. Scientifically it is also full of surprises. It is a very fast-moving cluster, at 490 km per second, the fastest known of any similar object. It also appears to be moving around the Milky Way in a direction opposite to most other objects,[10] which does suggest that it may be a captured globular cluster or even, at one time, part of another dwarf galaxy. Finally, although its declination is very far south, it is, believe it or not, visible from a latitude equivalent to that of London, although it will be skimming the horizon. Now that is an observing challenge!

NGC 3228	Collinder 218	10^h 21.4^m	−51° 43′	OC
6 m	⊕ 5′	20	II 3 p	Easy

Another nice object that can be glimpsed in a finderscope and easily located with binoculars. It is a nice group of about 8 or 9 stars ranging in magnitude from 7 to 10, set among a background of fainter stars. Increasing aperture resolves more members.

NGC 3330	Collinder 226	10^h 38.8^m	−54° 07′	OC
7.4 m	⊕ 6′	35	III 2 m	Easy

Our final object in Vela is just about visible in a finder and quite easy to locate in binoculars. It is a loose and irregular object with around two dozen stars in a rough north to south aspect. Increasing aperture will show more of the cluster's fainter members. Not a particularly spectacular cluster, but nevertheless a nice one to close our visit to Vela.

Notes

[10]This is known as retrograde motion and applies to any celestial object moving in a direction contrary to most other objects.

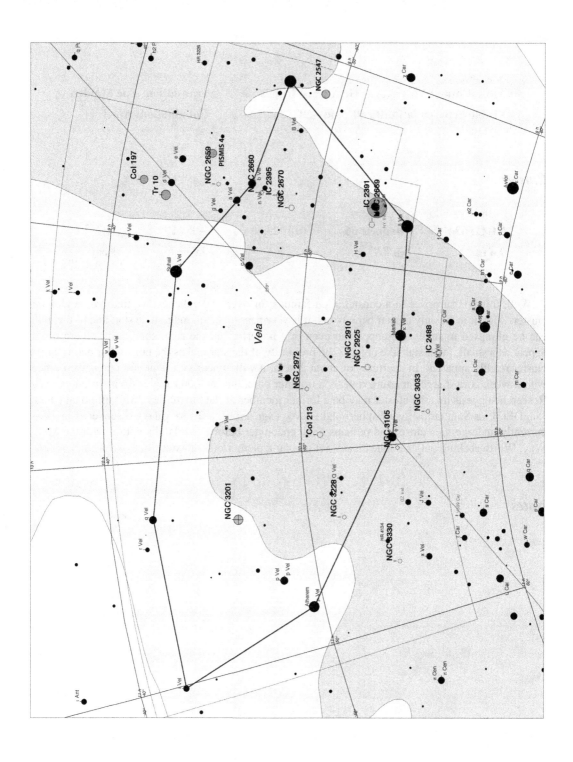

Virgo

Fast Facts

Abbreviation: Vir	**Genitive: Virginis**	**Translation: The Maiden**
Visible between latitudes 80° and −80°		**Culmination: April**

Cluster

NGC 5634	**Bennett 66**	**14^h 29.6^m**	**−05° 59′**	**GC**
9.4 m	**⊕ 4.9′**		**IV**	**Easy**

Well, this is a surprise! In a constellation famous for over 250 galaxies, we find a nice globular cluster. Alas, it is the only cluster hereabouts. It is very near to an 8th magnitude star, and in fact both can be glimpsed in small telescopes and even in a finder, given the right conditions, where it will appear as a small, faint, star-like object. The proximity of the star makes the non-stellar nature of the cluster readily apparent. In apertures of about 20 cm, it will appear as a faint and unresolved object with a weak core. Larger aperture reveals a brighter core, but the outer regions remain unresolved. Research suggests that the cluster may be a former member of the Sagittarius dwarf elliptical galaxy (SagDEG), or Sagittarius dwarf spheroidal galaxy (Sgr dSph), a dwarf galaxy discovered in 1994, currently undergoing a close, and perhaps final, encounter before its tidal disruption with the Milky Way. Worth seeking out as a lone cluster set among a plethora of galaxies.

Notes

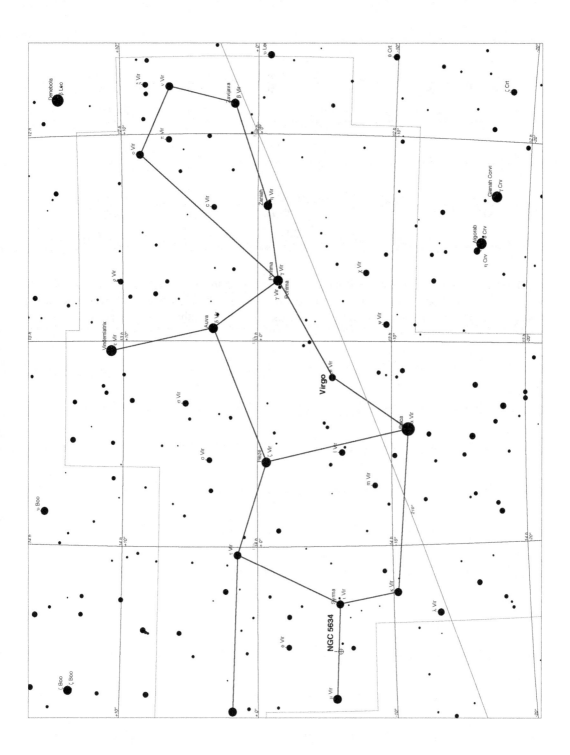

Volans

Fast Facts

Abbreviation: Vol	**Genitive: Volantis**	**Translation: The Flying Fish**
Visible between latitudes 10° and −90°		**Culmination: January**

Cluster

NGC 2348	–	07ʰ 03.0ᵐ	−67° 24′	OC
−m	⊕ 11′	30	–	Moderate

This solitary cluster is an irregularly shaped open cluster of about 30 stars when seen in a telescope of aperture 20 cm. There is a nice 10th magnitude white star south of its center. However, when using a larger aperture and a higher magnification the cluster stands out rather well because it is located in a relatively starless patch of sky.

Notes

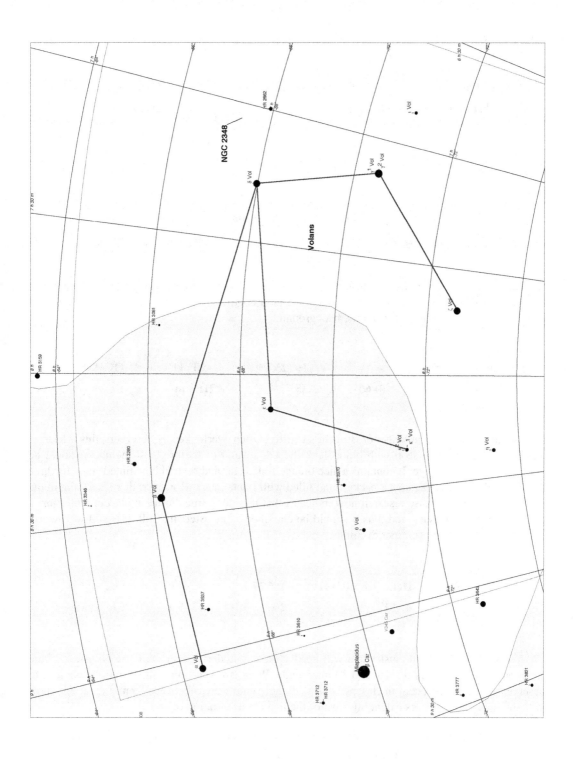

Vulpecula

Fast Facts

Abbreviation: Vul	**Genitive: Vulpeculae**	**Translation: The Fox**
Visible between latitudes 90° and −55°		**Culmination: July**

Clusters

NGC 6793	Herschel VIII-81	19^h 23.2^m	+22° 08′	OC
-m	⊕ 6′	215	III 2 p	Moderate

Located in a rich star field, the cluster consists of around forty 10th magnitude stars and fainter. It is not a particularly conspicuous cluster, and under poor conditions, and small to medium aperture, it may be difficult to distinguish from the background.

Collinder 399	–	19^h 25.4^m	+20° 11′	OC/Ast
3.6 m	⊕ 60′	35	III 3 m	Easy

Also known as the Coathanger or Brocchi's Cluster. Often overlooked by observers, this is a large, dissipated cluster easily seen with binoculars; indeed, several of the brightest members should be visible with the naked eye. It contains a nice orange-tinted star and several blue tinted stars. Its three dozen members are set against a background filled with fainter stars. However, there is a sting in the tail of this cluster, because research now indicates that it is not a true cluster at all, but rather just a chance alignment of stars, and thus it should be classified as an asterism. Still, it is still well worth observing during warm summer evenings.

NGC 6800	Herschel VIII-21	19^h 27.1^m	+25° 08′	OC
−m	⊕ 15′	20	IV 1 p	Moderate

This is a large but faint cluster that will need at least a moderate aperture to be observed. There are about two dozen stars of 12th and 13 magnitude. What is interesting is that it will be obvious to observers that the stars occur in clumps; with seemingly empty areas between them. Larger apertures may show two arcs of stars extending west of the main concentration of stars.

NGC 6802	Herschel VI-14	19^h 30.6^m	+20° 16′	OC
8.8 m	⊕ 5.0′	50	I 1 m	Moderate

This cluster can be a problem to find, sometimes. It all depends on the conditions, and if there is light pollution, such as at a typical urban location, then it will all but disappear. In small apertures it will appear as a very faint, small ball of light. Of course, larger apertures will reveal more resolvable stars. In order to help locate the cluster it is immediately to the east of the star 7 Vulpeculae, which is the most easterly star of the Coathanger asterism.

Stock 1	–	19ʰ 35.8ᵐ	+25° 13′	OC
5.3 m	⊕ 60′	40	III 2 m	Easy

An enormous cluster best seen in binoculars, although it is difficult to estimate where the cluster ends and the background stars begin. Strangely, this object is not often observed, both at amateur and professional levels. Apparently, initial research indicates it is a very young cluster.

NGC 6823	Herschel VII-18	19ʰ 43.2ᵐ	+23° 18′	OC
7.1 m	⊕ 7.0′	30	I 3 m n	Easy

Although this is a small cluster, it is quite a rich cluster. Consisting of about 30 or more stars concentrated in a small area of only 6′. Keen-eyed observers will see the stars forming a distinct 5′ oval ring, elongated approximately east to west. Under perfect conditions, you will see that several stars have delicate tints of yellow, orange and blue. Finally, the complete cluster is immersed in the nebula NGC 6820 that will need a filter to be seen.

NGC 6830	Herschel VII-9	19ʰ 51.0ᵐ	+23° 06′	OC
7.9 m	⊕ 12′	20	II 2 p	Easy

Standing out quite well, we have a nice open cluster. It is, admittedly, faint and small, larger apertures need to be used to appreciate it fully. However, having said that, it does stand out quite well in a low aperture with low magnification. With a large aperture, many double stars can be seen, as well as small clumps.

NGC 6834	Herschel VIII-16	19ʰ 52.2ᵐ	+29° 24′	OC
7.8 m	⊕ 5′	50	II 2 m	Moderate

A faint but rich object of about four dozen stars ranging in magnitude from 10 to the very faint 15. With a small aperture it will only appear very small, somewhat elongated from west to east. This is where averted vision comes in handy, as, with it, many fainter stars are resolved, especially around its center, making the cluster appear circular.

NGC 6885	Herschel VIII-20	20^h 12.0^m	+26° 29′	OC
5.9 m	⊕ 7′	25	III 2 p	Easy

This cluster, also known as Caldwell 37, is an irregular object containing many 9th to 13th magnitude stars. This is one of a number of objects that apparently were unknown to the early amateur astronomer. Visible in binoculars as a hazy blur, it may look as if it is located next to the cluster NGC 6882 and thus is not easily delineated. However, it is now believed that they are one and the same thing. This is an old cluster, with an estimated age of around 1 billion years, with recent measurements placing it at a distance of 1,200 light years.

NGC 6940	Herschel VII-8	20^h 34.6^m	+28° 18′	OC
6.3 m	⊕ 32′	75	III 2 r	Easy

A beautiful cluster, which although visible in binoculars, is best appreciated with a small aperture telescope. It contains the semi-variable star FG Vulpeculae near its center that has a nice reddish-orange tint. Set against the rich panoply of the Milky Way, you can spend a long time observing this part of the sky. And if you are lucky enough to have a medium aperture telescope, then it really is spectacular, with double stars, arcs and chains of stars, and many areas devoid of stars. A delightful object with which to leave Vulpecula and end our journey!

Notes

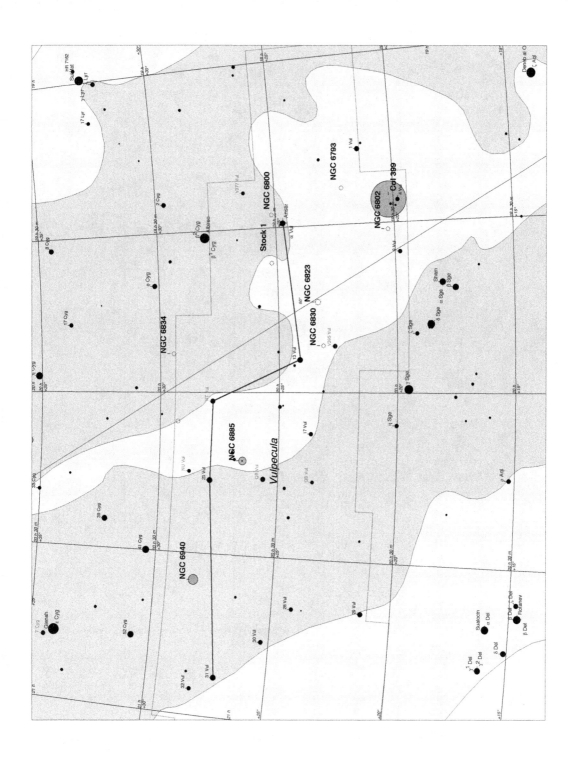

Appendix: Books, Magazines, and Organizations

There are many fine astronomy books in print, and to choose among them is a difficult task. Nevertheless there are a few that are among the best on offer. (Author: need pub dates)

Norton's Star Atlas & Reference Handbook, I. Ridpath (Ed.)
Sky Atlas 2000.0, W. Tirion, R. Sinnott, Sky Publishing & Cambridge University Press
Millennium Star Atlas, R. Sinnott, M. Perryman, Sky Publishing
Astronomy of the Milky Way, Volume I – Observer's Guide to the Northern Sky, M. Inglis, Springer
Astronomy of the Milky Way, Volume II – Observer's Guide to the Southern Sky, M. Inglis, Springer
A Field Guide to the Deep-Sky Objects, 2nd Edition, M. Inglis, Springer
Astrophysics Is Easy, M. Inglis, Springer
The Caldwell Objects and How To Observe Them, M. Mobberley, Springer
The Herschel Objects and How To Observe Them, J. Mullaney, Springer
Star Clusters and How To Observe Them, M. Allison, Springer

The first three magazines are aimed at a general audience and so are applicable to everyone; the last three are aimed at the well-informed layperson. In addition there are many research level journals that can be found in university libraries and observatories.

Astr-onomy Now
Sky & Telescope
New Scientist
Scientific American
Science
Nature

M. Inglis, *Observer's Guide to Star Clusters*, The Patrick Moore Practical Astronomy Series,   273
DOI 10.1007/978-1-4614-7567-5, © Springer Science+Business Media New York 2013

Organizations

The Federation of Astronomical Societies

[http://www.fedastro.org.uk/fas/]
Society for Popular Astronomy
[http://www.popastro.com/]
The British Astronomical Association
[http://britastro.org/baa/]
The Royal Astronomical Society
[http://www.ras.org.uk/membership.htm]
Campaign for Dark Skies
[http://www.britastro.org/dark-skies/]
International Dark-Sky Association
[http://www.darksky.org/]
The Astronomical League
[http://www.astroleague.org/]
The Webb Deep-Sky Society
[http://www.webbdeepsky.com/]

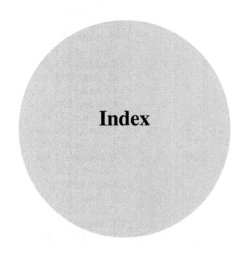

Index

B
Basel 4, 31
Bennett 66, 264
Berkeley 68, 187
Blanco 1, 230
Bochum 1, 115
Bochum 7, 259
Bochum 13, 224
Burnham 584, 40

C
Caldwell 1, 75
Caldwell 4, 72
Caldwell 13, 61
Caldwell 14, 2, 184
Caldwell 25, 142
Caldwell 41, 240
Caldwell 42, 102
Caldwell 47, 102
Caldwell 54, 156
Caldwell 58, 45
Caldwell 64, 46
Caldwell 66, 124
Caldwell 71, 2, 195
Caldwell 73, 82
Caldwell 78, 88
Caldwell 79, 261
Caldwell 80, 67
Caldwell 84, 68
Caldwell 85, 258
Caldwell 87, 122
Caldwell 88, 78
Caldwell 89, 163
Caldwell 91, 55
Caldwell 93, 180

Caldwell 94, 3, 92
Caldwell 96, 54
Caldwell 97, 66
Caldwell 98, 91
Caldwell 100, 66
Caldwell 104, 250
Caldwell 105, 159
Caldwell 106, 250
Caldwell 108, 158
Collinder 3, 59
Collinder 5, 60
Collinder 6, 74
Collinder 12, 60
Collinder 15, 61
Collinder 21, 248
Collinder 22, 182
Collinder 24, 184
Collinder 28, 184
Collinder 29, 185
Collinder 31, 185
Collinder 33, 14, 63
Collinder 34, 63
Collinder 37, 176
Collinder 38, 177, 225
Collinder 39, 185
Collinder 40, 185
Collinder 41, 186
Collinder 44, 186
Collinder 51, 187
Collinder 55, 176
Collinder 65, 241
Collinder 67, 31
Collinder 69, 176
Collinder 70, 178
Collinder 73, 176
Collinder 75, 31

M. Inglis, *Observer's Guide to Star Clusters*, The Patrick Moore Practical Astronomy Series,   275
DOI 10.1007/978-1-4614-7567-5, © Springer Science+Business Media New York 2013

Collinder 76, 177
Collinder 77, 114
Collinder 79, 177
Collinder 80, 114
Collinder 81, 114
Collinder 82, 114
Collinder 87, 177
Collinder 89, 115
Collinder 91, 150
Collinder 92, 150
Collinder 94, 151
Collinder 96, 151
Collinder 97, 151
Collinder 100, 151
Collinder 102, 152
Collinder 103, 152
Collinder 104, 152
Collinder 106, 152
Collinder 107, 152, 153
Collinder 111, 153
Collinder 114, 150
Collinder 115, 153
Collinder 118, 44
Collinder 121, 44
Collinder 124, 154
Collinder 126, 115
Collinder 132, 45
Collinder 134, 45
Collinder 135, 192
Collinder 136, 45
Collinder 137, 46
Collinder 139, 47
Collinder 140, 46
Collinder 144, 116
Collinder 147, 192
Collinder 150, 193
Collinder 155, 193
Collinder 158, 194
Collinder 159, 194
Collinder 161, 195
Collinder 162, 195
Collinder 164, 195
Collinder 165, 195
Collinder 170, 155
Collinder 172, 54
Collinder 173, 197
Collinder 175, 197
Collinder 177, 258
Collinder 178, 198
Collinder 179, 124
Collinder 182, 198
Collinder 189, 40
Collinder 190, 202
Collinder 191, 258
Collinder 193, 259
Collinder 195, 202
Collinder 197, 259
Collinder 200, 259
Collinder 202, 260
Collinder 203, 260

Collinder 204, 40
Collinder 206, 202
Collinder 208, 260
Collinder 209, 260
Collinder 211, 261
Collinder 213, 261
Collinder 214, 261
Collinder 218, 262
Collinder 226, 262
Collinder 238, 55
Collinder 247, 66
Collinder 249, 66
Collinder 251, 90
Collinder 252, 90
Collinder 254, 90
Collinder 255, 91
Collinder 259, 91
Collinder 260, 158
Collinder 263, 91
Collinder 264, 91
Collinder 265, 158
Collinder 270, 67
Collinder 272, 67
Collinder 277, 159
Collinder 279, 68
Collinder 280, 68
Collinder 283, 69
Collinder 284, 69
Collinder 286, 78
Collinder 287, 138
Collinder 289, 138
Collinder 290, 78
Collinder 291, 162
Collinder 292, 162
Collinder 293, 162
Collinder 296, 118
Collinder 297, 162
Collinder 298, 163
Collinder 300, 163
Collinder 301, 220
Collinder 303, 163
Collinder 304, 164
Collinder 306, 164
Collinder 307, 26
Collinder 308, 221
Collinder 309, 222
Collinder 315, 222
Collinder 316, 222, 223
Collinder 317, 222
Collinder 319, 223
Collinder 322, 223
Collinder 323, 223
Collinder 326, 224
Collinder 332, 224
Collinder 335, 224
Collinder 336, 224
Collinder 338, 225
Collinder 339, 225
Collinder 341, 226
Collinder 342, 226

Collinder 344, 226, 227
Collinder 350, 172
Collinder 354, 227
Collinder 356, 210
Collinder 363, 211
Collinder 367, 212
Collinder 371, 213
Collinder 373, 236
Collinder 375, 236
Collinder 376, 213
Collinder 378, 215
Collinder 382, 214
Collinder 384, 232
Collinder 386, 237
Collinder 389, 233
Collinder 390, 233
Collinder 391, 233
Collinder 392, 22
Collinder 394, 215
Collinder 399, 268
Collinder 402, 94
Collinder 403, 94
Collinder 413, 94
Collinder 415, 95
Collinder 419, 95
Collinder 422, 95
Collinder 425, 96
Collinder 426, 18
Collinder 428, 96
Collinder 429, 72
Collinder 430, 96
Collinder 431, 97
Collinder 432, 96
Collinder 434, 97
Collinder 435, 97
Collinder 436, 97
Collinder 438, 98
Collinder 443, 73
Collinder 445, 130
Collinder 446, 73
Collinder 447, 73
Collinder 450, 73
Collinder 455, 58
Collinder 463, 62
Collinder 464, 37
Collinder 470, 98
Czernik 39, 23

D
Dolidze 26, 50
Dunlop 273, 68
Dunlop 366, 28

H
Haffner 6, 46
Haffner 8, 46
Harvard 12, 223
Harvard 15, 171

Harvard 20, 206
Herschel 69, 14
Herschel 72, 172
Herschel I-19, 84
Herschel I-44, 172
Herschel I-45, 170
Herschel I-46, 171
Herschel I-47, 233
Herschel I-48, 171
Herschel I-49, 210
Herschel I-50, 213
Herschel I-147, 170
Herschel I-149, 171
Herschel II-195, 170
Herschel II-197, 211
Herschel II-200, 211
Herschel II-201, 212
Herschel II-584, 168
Herschel IV-12, 212
Herschel IV-50, 118
Herschel V-558, 19
Herschel VI-1, 116
Herschel VI-2, 115
Herschel VI-6, 116
Herschel VI-7, 85
Herschel VI-9, 34
Herschel VI-10, 221
Herschel VI-11, 169
Herschel VI-12, 170
Herschel VI-14, 268
Herschel VI-19, 136
Herschel VI-20, 230
Herschel VI-21, 115
Herschel VI-23, 215
Herschel VI-25, 185
Herschel VI-27, 154
Herschel VI-30, 58
Herschel VI-31, 62
Herschel VI-32, 98
Herschel VI-35, 59
Herschel VI-39, 198
Herschel VI-40, 168
Herschel VII-7, 210
Herschel VII-8, 270
Herschel VII-9, 269
Herschel VII-10, 196
Herschel VII-11, 197
Herschel VII-13, 44
Herschel VII-18, 269
Herschel VII-20, 150
Herschel VII-23, 196
Herschel VII-28, 194
Herschel VII-30, 212
Herschel VII-33, 30
Herschel VII-38, 154
Herschel VII-40, 95
Herschel VII-41, 131
Herschel VII-45, 60
Herschel VII-46, 61
Herschel VII-47, 36

Herschel VII-48, 61
Herschel VII-49, 61
Herschel VII-53, 130
Herschel VII-56, 58
Herschel VII-58, 196
Herschel VII-59, 94
Herschel VII-60, 186
Herschel VII-61, 187
Herschel VII-63, 202
Herschel VII-64, 198
Herschel VII-66, 72
Herschel VII-67, 193
Herschel VIII-1, 196
Herschel VIII-3, 152
Herschel VIII-5, 153
Herschel VIII-12, 232
Herschel VIII-16, 269
Herschel VIII-20, 270
Herschel VIII-21, 268
Herschel VIII-25, 151
Herschel VIII-29, 130
Herschel VIII-30, 197
Herschel VIII-31, 153
Herschel VIII-32, 155
Herschel VIII-33, 155
Herschel VIII-34, 155
Herschel VIII-39, 154
Herschel VIII-56, 95
Herschel VIII-59, 30
Herschel VIII-61, 30
Herschel VIII-64, 60
Herschel VIII-65, 62
Herschel VIII-66, 62
Herschel VIII-68, 32
Herschel VIII-75, 130
Herschel VIII-78, 60
Herschel VIII-79, 59
Herschel VIII-80, 186
Herschel VIII-81, 268
Herschel VIII-85, 187
Herschel VIII-88, 186
Hodge 3, 112
Hogg 22, 26, 27
Hyades Stream, 242

I
IC 348, 186
IC 1369, 96
IC 1396, 72
IC 1434, 130
IC 1442, 131
IC 2157, 114
IC 2391, 258
IC 2395, 259
IC 2488, 260
IC 2602, 55
IC 2714, 56
IC 2944, 66

IC 4499, 16
IC 4550, 162
IC 4665, 172
IC 4725, 214
IC 4756, 172, 237
IC 5146, 98, 99

K
King 12, 58
King 14, 59
King 19, 74

L
Lambda Orionis Association, 176
Lund 57, 62
Lund 709, 26
Lund 940, 95
Lund 992, 97, 98
Lund 1049, 182

M
Melotte 20, 185
Melotte 22, 240
Melotte 25, 240
Melotte 28, 241
Melotte 29, 241
Melotte 66, 192
Melotte 71, 193
Melotte 111, 84
Melotte 169, 27
Melotte 227, 166
Messier 2, 18
Messier 3, 8, 42
Messier 4, 28, 220
Messier 5, 236
Messier 9, 170, 171
Messier 10, 168, 169
Messier 12, 168, 169
Messier 13, 42, 118, 119, 215
Messier 14, 172
Messier 15, 182
Messier 18, 168, 213
Messier 19, 169
Messier 21, 18, 211
Messier 22, 215
Messier 23, 210
Messier 24, 212
Messier 25, 115, 215
Messier 28, 214
Messier 29, 2, 95
Messier 30, 52
Messier 34, 185
Messier 35, 114
Messier 36, 31
Messier 37, 32
Messier 38, 31

Messier 39, 98
Messier 41, 3, 45
Messier 44, 2, 40, 242
Messier 45, 240
Messier 46, 194
Messier 47, 193, 194
Messier 48, 124
Messier 50, 154
Messier 52, 58
Messier 53, 84, 85
Messier 54, 134, 216
Messier 55, 216
Messier 56, 144
Messier 62, 169
Messier 67, 40
Messier 68, 124
Messier 69, 214, 215
Messier 70, 215
Messier 71, 206
Messier 72, 18
Messier 73, 18
Messier 75, 216
Messier 79, 134
Messier 80, 220
Messier 92, 118
Messier 93, 195
Messier 103, 61

N
NGC 103, 59
NGC 104, 250
NGC 129, 59
NGC 133, 59
NGC 136, 59
NGC 146, 60
NGC 188, 74
NGC 225, 60
NGC 288, 230
NGC 330, 250, 262
NGC 362, 250
NGC 381, 60
NGC 436, 60
NGC 457, 60
NGC 559, 61
NGC 581, 61
NGC 637, 61, 234
NGC 643, 128
NGC 654, 61
NGC 659, 62
NGC 663, 62
NGC 744, 184
NGC 752, 14
NGC 869, 184
NGC 884, 184
NGC 957, 184
NGC 1027, 62
NGC 1039, 185
NGC 1049, 112

NGC 1245, 185
NGC 1261, 122
NGC 1342, 186
NGC 1444, 186
NGC 1466, 128
NGC 1496, 186
NGC 1502, 36
NGC 1513, 186
NGC 1528, 187
NGC 1545, 187
NGC 1582, 187
NGC 1647, 240
NGC 1662, 176
NGC 1664, 30
NGC 1731, 106
NGC 1746, 241
NGC 1755, 106
NGC 1761, 106
NGC 1778, 30
NGC 1807, 241
NGC 1817, 241
NGC 1818, 107
NGC 1820, 107
NGC 1848, 148
NGC 1850, 107
NGC 1851, 82, 192
NGC 1857, 30
NGC 1858, 107
NGC 1866, 108
NGC 1893, 30
NGC 1904, 134, 192
NGC 1912, 31
NGC 1955, 108
NGC 1960, 31
NGC 1968, 108
NGC 1974, 108
NGC 1980, 176
NGC 1981, 176
NGC 1983, 108
NGC 1984, 108
NGC 1994, 108
NGC 2004, 109
NGC 2017, 134
NGC 2042, 109
NGC 2050, 109
NGC 2055, 109
NGC 2099, 31
NGC 2100, 109
NGC 2112, 177
NGC 2126, 32
NGC 2129, 114
NGC 2132, 190
NGC 2141, 177
NGC 2158, 114
NGC 2168, 114
NGC 2169, 177
NGC 2194, 177
NGC 2204, 44
NGC 2215, 150

NGC 2232, 151
NGC 2236, 151
NGC 2243, 44
NGC 2250, 151
NGC 2251, 152
NGC 2252, 152
NGC 2254, 152
NGC 2264, 153
NGC 2266, 115
NGC 2269, 150
NGC 2281, 32
NGC 2286, 153
NGC 2287, 44
NGC 2298, 192
NGC 2301, 154
NGC 2302, 154
NGC 2304, 115
NGC 2323, 154
NGC 2324, 154
NGC 2331, 115
NGC 2335, 155
NGC 2343, 155
NGC 2348, 266
NGC 2353, 155
NGC 2354, 45
NGC 2355, 45
NGC 2360, 45
NGC 2362, 45
NGC 2367, 46
NGC 2374, 47
NGC 2383, 47, 48
NGC 2384, 47, 48
NGC 2394, 50
NGC 2395, 116
NGC 2414, 193
NGC 2419, 142
NGC 2420, 116
NGC 2421, 193
NGC 2422, 193
NGC 2423, 194
NGC 2437, 194
NGC 2439, 194
NGC 2447, 194
NGC 2451, 195
NGC 2453, 195
NGC 2467, 195
NGC 2477, 195
NGC 2479, 196
NGC 2482, 196
NGC 2489, 196
NGC 2506, 155
NGC 2509, 196
NGC 2516, 47, 54
NGC 2527, 197
NGC 2533, 197
NGC 2539, 197
NGC 2546, 198
NGC 2547, 47, 258
NGC 2548, 124

NGC 2567, 198
NGC 2571, 198
NGC 2579, 198
NGC 2587, 199
NGC 2627, 202
NGC 2632, 40
NGC 2635, 202
NGC 2658, 202
NGC 2660, 259
NGC 2669, 260
NGC 2670, 259
NGC 2682, 40
NGC 2808, 54, 192
NGC 2818, 202, 203
NGC 2910, 260
NGC 2972, 261
NGC 2999, 261
NGC 3105, 261
NGC 3114, 54, 55
NGC 3201, 261
NGC 3228, 262
NGC 3293, 55
NGC 3330, 262
NGC 3532, 55
NGC 3603, 56
NGC 3680, 66
NGC 3766, 66
NGC 4052, 90
NGC 4103, 90
NGC 4147, 84, 85
NGC 4337, 90
NGC 4349, 91
NGC 4372, 158
NGC 4439, 91
NGC 4463, 158
NGC 4590, 124
NGC 4609, 91
NGC 4755, 91
NGC 4815, 158
NGC 4833, 159
NGC 5024, 84
NGC 5053, 85
NGC 5138, 67
NGC 5139, 67
NGC 5272, 8, 42
NGC 5281, 68
NGC 5286, 68
NGC 5316, 68
NGC 5385, 256
NGC 5460, 68
NGC 5466, 34
NGC 5617, 69
NGC 5634, 264
NGC 5662, 69
NGC 5694, 124
NGC 5715, 78
NGC 5749, 138
NGC 5822, 78, 138
NGC 5823, 78, 79

NGC 5824, 138
NGC 5897, 136
NGC 5904, 236
NGC 5925, 162
NGC 5927, 138
NGC 5946, 162
NGC 5986, 139
NGC 5999, 162
NGC 6025, 246
NGC 6031, 162
NGC 6067, 163
NGC 6087, 163
NGC 6093, 220
NGC 6101, 16
NGC 6115, 163
NGC 6121, 220
NGC 6124, 220
NGC 6134, 163
NGC 6139, 221
NGC 6144, 221
NGC 6152, 164
NGC 6169, 164
NGC 6171, 168
NGC 6178, 221
NGC 6188, 26
NGC 6192, 222
NGC 6193, 26
NGC 6200, 26
NGC 6204, 27
NGC 6205, 118
NGC 6208, 27
NGC 6218, 168
NGC 6229, 118
NGC 6231, 222
NGC 6235, 168
NGC 6242, 222
NGC 6249, 223
NGC 6250, 27
NGC 6254, 169
NGC 6259, 223
NGC 6266, 169
NGC 6268, 223
NGC 6273, 169
NGC 6284, 169
NGC 6287, 170
NGC 6293, 170
NGC 6304, 170
NGC 6316, 170
NGC 6322, 224
NGC 6333, 170
NGC 6341, 118
NGC 6342, 171
NGC 6355, 171
NGC 6356, 171
NGC 6362, 27
NGC 6366, 171
NGC 6383, 224, 225
NGC 6388, 225

NGC 6396, 225
NGC 6397, 28, 220
NGC 6400, 226
NGC 6401, 172
NGC 6402, 172
NGC 6405, 226
NGC 6416, 226
NGC 6425, 227
NGC 6441, 227
NGC 6469, 210
NGC 6475, 227
NGC 6494, 210
NGC 6496, 228
NGC 6520, 210
NGC 6522, 210
NGC 6528, 211
NGC 6530, 211
NGC 6531, 211
NGC 6541, 88
NGC 6544, 211
NGC 6546, 211
NGC 6553, 212
NGC 6568, 212
NGC 6569, 212
NGC 6584, 244
NGC 6595, 213
NGC 6604, 236
NGC 6611, 236
NGC 6613, 213
NGC 6618, 213
NGC 6624, 213, 228
NGC 6626, 214
NGC 6633, 172
NGC 6637, 214, 228
NGC 6638, 214
NGC 6645, 215
NGC 6649, 232
NGC 6652, 214
NGC 6656, 215
NGC 6664, 232
NGC 6681, 215
NGC 6683, 232
NGC 6694, 233
NGC 6704, 233
NGC 6705, 233
NGC 6709, 22
NGC 6712, 233
NGC 6715, 216
NGC 6716, 215
NGC 6723, 216
NGC 6738, 22
NGC 6749, 22
NGC 6752, 180
NGC 6755, 22, 23
NGC 6756, 23
NGC 6760, 23
NGC 6773, 23
NGC 6775, 23

NGC 6779, 144
NGC 6791, 144
NGC 6793, 268
NGC 6795, 23
NGC 6800, 268
NGC 6802, 268
NGC 6809, 216
NGC 6811, 94
NGC 6819, 94
NGC 6823, 269
NGC 6828, 23
NGC 6830, 269
NGC 6834, 269
NGC 6838, 206
NGC 6864, 216
NGC 6866, 94
NGC 6871, 94
NGC 6873, 206
NGC 6883, 95
NGC 6885, 270
NGC 6910, 95
NGC 6913, 95
NGC 6934, 34, 102
NGC 6939, 72
NGC 6940, 270
NGC 6950, 102
NGC 6981, 18
NGC 6994, 18
NGC 6996, 96
NGC 7006, 102
NGC 7023, 72
NGC 7031, 96
NGC 7039, 97
NGC 7062, 97
NGC 7063, 97
NGC 7067, 97
NGC 7078, 182
NGC 7082, 97, 98
NGC 7086, 98
NGC 7089, 18
NGC 7092, 98
NGC 7099, 52
NGC 7128, 95
NGC 7142, 72
NGC 7160, 73
NGC 7209, 130
NGC 7226, 73
NGC 7235, 73
NGC 7243, 130
NGC 7245, 130
NGC 7261, 73
NGC 7296, 131
NGC 7380, 74
NGC 7492, 19
NGC 7510, 74
NGC 7654, 58
NGC 7686, 14
NGC 7748, 74
NGC 7772, 182

NGC 7789, 58, 223
NGC 7790, 58

O
Orion Association, 178, 228

P
Palomar 10, 206
Palomar 12, 52
Palomer 11, 23
Pismis 4, 258
Praesepe, 40, 242

R
Ruprecht 18, 47
Ruprecht 20, 48
Ruprecht 98
Ruprecht 118, 163

S
Scorpius-Centaurus Association, 5, 228
Stephenson 1, 144
Stock 1, 269
Stock 2, 62
Stock 14, 66
Stock 23, 36

T
Tombaugh 5, 36
Trumpler 1, 61
Trumpler 2, 185
Trumpler 10, 260
Trumpler 21, 67
Trumpler 22, 69
Trumpler 24, 223
Trumpler 26, 171
Trumpler 27, 224, 225
Trumpler 28, 225
Trumpler 29, 226
Trumpler 30, 227
Trumpler 33, 215
Trumpler 34, 232
47 Tucanae, 250

U
Upgren 1, 42
Ursa Major Stream, 5, 242, 254

Z
Zeta Persei Association, 186, 188